Mohamed Elzagheid
Chemical Technicians

Also of Interest

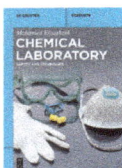

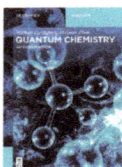

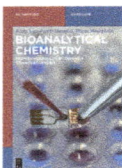

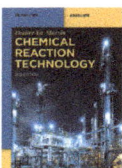

Mohamed Elzagheid

Chemical Technicians

Good Laboratory Practice and Laboratory Information
Management Systems

DE GRUYTER

Author
Prof. Dr. Mohamed Elzagheid
Chemical Engineering Department
Jubail Industrial College
Jubail Industrial City
Jubail 31961
Saudi Arabia
elzagheid_m@rcjy.edu.sa; melzagheid@gmail.com

ISBN 978-3-11-119110-2
e-ISBN (PDF) 978-3-11-119149-2
e-ISBN (EPUB) 978-3-11-119162-1

Library of Congress Control Number: 2023939547

Bibliographic information published by the Deutsche Nationalbibliothek
The Deutsche Nationalbibliothek lists this publication in the Deutsche Nationalbibliografie;
detailed bibliographic data are available on the Internet at http://dnb.dnb.de.

© 2023 Walter de Gruyter GmbH, Berlin/Boston
Cover image: Nguyen Dinh Minh Quan/iStock/Getty Images Plus
Typesetting: Integra Software Services Pvt. Ltd.
Printing and binding: CPI books GmbH, Leck

www.degruyter.com

Preface

This chemical technician-specific book describes a variety of skills that chemical technicians and chemical plant operations technicians should acquire as part of their valuable experience. The book presents these competencies in a sequential manner that is easy to follow and understand, which were unintentionally addressed in other resources in a scattered manner between internet resources and chapters in different textbooks.

The book provides a brief overview of the roles and responsibilities of various chemical laboratory technicians, as well as the tasks they must complete. It also includes a detailed description of the sampling procedures and chemical analyses, as well as a description of the various laboratory equipment, techniques, and safety procedures used in chemical laboratories and chemical plants.

The book also discusses information management systems and good laboratory practices, as well as how they have enabled and facilitated laboratory practices and the collection of information that enhances technicians' experience and knowledge.

Finally, some advice on using lab glassware, laboratory emergency first aid, and a short description of the chemicals that chemical technicians frequently use are provided.

Dr. Mohamed Ibrahim Elzagheid, Chemistry Professor

Waterloo, Ontario, Canada

2023

https://doi.org/10.1515/9783111191492-202

Acknowledgment

This book is dedicated to all those who have supported and encouraged me in pursuing writing and sharing my expertise.

First and foremost, I would like to express my heartfelt gratitude to my entire family for always assisting and supporting me throughout my academic career, whether it is teaching, research, or writing.

I would also like to thank my colleagues in the chemical engineering department for the fruitful discussion and the excellent ideas they always share. Special thanks to Majid Al Hazza, who assisted in the early stages of writing this book by providing useful material.

Last but not least, a special thanks to the entire publishing team, especially Ute Skambraks and Christene Smith, whose support and work cannot go unmentioned.

https://doi.org/10.1515/9783111191492-203

The Author

Mohamed Elzagheid is an Associate Professor of Chemistry at Jubail Industrial College (JIC), which serves the Royal Commission for Jubail and Yanbu in Saudi Arabia. In addition, he works as a professor and consultant for the Libyan Authority for Scientific Research, which is part of the Ministry of Education in Libya.

During his 23-year career at McGill University, SynPrep Inc. in Montreal, Canada, and Jubail Industrial College in Saudi Arabia, he was directly involved in the education of laboratory technicians and chemists, as well as supervised many undergraduate and graduate chemistry students, and has significantly contributed to numerous training programs for Saudi Arabian and international companies, both short-term and long-term, for chemistry-based laboratory technicians.

He also served on the Research, Publications, Projects, and Academic Promotion Team; the Academic Promotion Committee; the Curriculum Development Committee; the Industrial Chemistry Technology Program Advisory and Evaluation Committee; the CTAB Steering Accreditation Committee; the Industrial Outreach Committee; and the Chemical Engineering Department Safety Committee at Jubail Industrial College.

Dr. Elzagheid has published four textbooks, which are currently being used to train chemistry-based technicians: two focused on organic chemistry, one in macromolecular chemistry, and one in chemical laboratory safety and techniques.

His research at Turku University in Finland, McGill University in Canada, and JIC in the Kingdom of Saudi Arabia has helped him establish a solid reputation in chemistry and chemical education in general, as evidenced by his research papers and publications.

https://doi.org/10.1515/9783111191492-204

Contents

Chapter 6
Chemical Laboratory and Chemical Plant Safety Procedures —— 60

Chapter 1
Introduction

1.1 Overview

In any industrial plant, pharmaceutical company, or manufacturing facility, many operations or processes can take place to change raw materials into desired products. In these workplaces, different workers with different backgrounds work together to ensure that all equipment, instruments, and laboratory experiments in the plant or the manufacturing facility work efficiently, run correctly, and are in excellent condition. Some of them are plant operators, and others are laboratory technicians based in the laboratory, undertaking various routine tasks and experiments. Those facilities may also have chemists and scientists or research and development researchers.

While the roles of chemists, scientists, and researchers can be clearly defined, the chemical laboratory technician's role in the industry is challenging to define because of the training, experience, and duties that can vary from one workplace to another. In specific chemical industries, company employees are recognized as chemical technicians simply because they work with chemicals. In contrast, other companies award this title to their employees only after a highly formalized education and years of on-the-job training.

1.2 Definitions

1.2.1 Chemical Laboratory Technician

Chemical laboratory technicians assist chemists and chemical engineers in the research, testing, chemical processes, and product development. They help design, run, and monitor experiments, and often record and report results. They have the required skills to execute laboratory techniques and instruments. They usually conduct experiments and monitor production processes on a full-time basis, and before joining work, they should have acquired a basic knowledge of chemistry, biochemistry, statistics, and other science subjects.

1.2.2 Chemical Plant Operation Technician

Chemical plant operators are responsible for running plant equipment safely and efficiently. As part of their job, they may need to transport materials, check equipment, monitor operating parameters, and keep records. They usually collaborate with technicians to handle and fix problems or do troubleshoot. In general, their main goal as

https://doi.org/10.1515/9783111191492-001

chemical plant operation technicians is to ensure that production is going smoothly without any difficulties.

1.2.3 Chemist

A chemist is a person who studies chemistry or works with chemicals or studies their reactions. He/she is a person who is trained in the chemistry field and studied the composition of chemical matter and its properties. He/she is also a person who does research connected with chemistry.

1.2.4 Scientist

A scientist is a person who has studied or acquired knowledge in one or more of the natural or physical sciences. He/she is also someone who systematically collects and uses research and evidence to formulate and test hypotheses, gain and share understanding and knowledge, or invent something that is useful and has applications. It can be further defined as someone with knowledge or expertise in any of the sciences, such as biology or chemistry.

1.2.5 Researcher

A researcher is a person who conducts academic or scientific research to discover new information or gain new understanding. He/she can also be considered a person whose goal is to study a subject in detail to discover new information or better understand the subject. A researcher is also someone who conducts systematic and organized research on something unknown.

1.2.6 Responsibility

This is the task that you are asked or expected to perform, or the state or fact of having the duty to deal with something, having control over someone, or being someone who causes something to happen. It can also be defined as the state of being responsible for something.

1.2.7 Skill

This is the ability to apply knowledge effectively and quickly in the performance of a task or the ability resulting from a person's knowledge, practice, and talent to do something well.

1.2.8 Duty

Behavior, services, or obligations arise from one's position. It can also be defined as an obligation, something one has to do as part of his/her job, or something that someone feels is the right thing to do.

1.2.9 Good Laboratory Practice (GLP)

An organizational process and the conditions under which nonclinical health and environmental research are planned, carried out, controlled, recorded, reported on, and kept or archived are covered by the management quality control system known as "Good Laboratory Practice," or GLP. It can also be described as the process used to generate, manage, report, maintain, and archive research data in support of assessments of the safety of people, animals, and the environment, as well as research targeted at advancing a risk-based approach to data management. Selected GLP principles are outlined in Figure 1.1 and will be covered in more detail in Chapter 7.

Figure 1.1: Selected Good Laboratory Practice principles.

1.2.10 Laboratory Information Management System (LIMS)

By utilizing the Laboratory Information Management System (LIMS), samples can be efficiently managed as well as the data that goes with them. LIMS can also be defined as a program that manages samples and associated data and can combine tools and automate procedures.

It is also known as a laboratory management system or an information system for laboratories. It can alternatively be described as a software program that supports the contemporary laboratory.

Modern laboratories require LIMS for a number of reasons, including sample monitoring, data management, workflow effectiveness, and scheduling equipment maintenance. The importance of purchasing and using a LIMS that is customized to their unique needs cannot be overstated because there is no one solution that works for all facilities. There are different LIMS choices that can be tailored to meet the requirements of different labs. The types of LIMS that are highlighted in Figure 1.2 will be discussed in greater detail in later chapters.

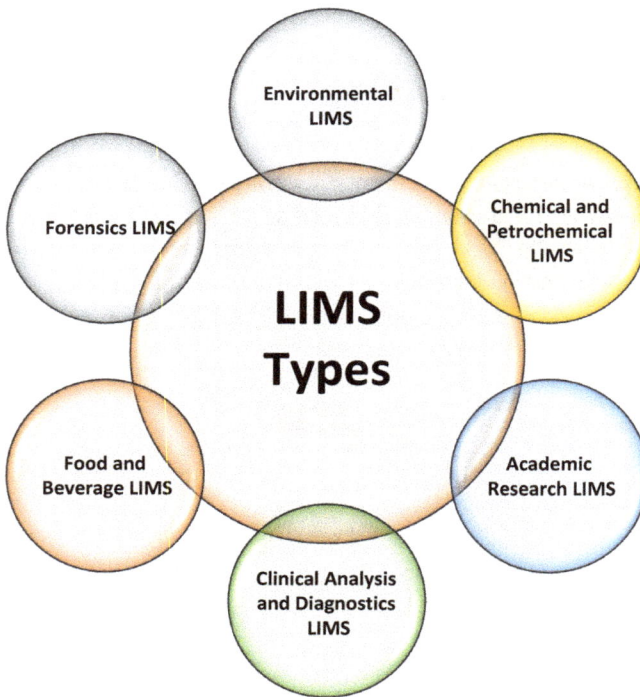

Figure 1.2: Selected LIMS types.

1.3 Questions

1.3.1 How many LIMS types are there?
1.3.2 What does LIMS stand for?
1.3.3 What is the difference between responsibility and duty?
1.3.4 List four GLP principles.
1.3.5 What for LIMS is required?

Chapter 2
Chemical Laboratory Technician Skills

To assist chemists, chemical technologists, and engineers with the research, development, production, and testing of chemical products and processes, chemical technicians typically use laboratory equipment and implement specific methodologies. They collaborate in teams, typically under the direction of chemists or chemical engineers who monitor, coach, and assess the outcomes. On the contrary, in some circumstances where chemical technicians have more expertise, they may train newly hired chemists in extremely specific fields of study. Chemical laboratory technicians could also help chemists and scientists by keeping an eye on some studies throughout the creation of new drugs. The knowledge gap between chemists and scientists might be reduced, thanks to them. Chemical technicians are employed in various fields, including testing laboratories, pharmaceutical and medicine manufacturing plants, colleges, universities, and professional schools, as shown in Figure 2.1.

Figure 2.1: Chemical technicians' workplaces.

Chemical technicians need to possess the following abilities, which are listed in brief on the following pages, in order to carry out their activities and responsibilities in various workplaces.

2.1 Knowledge or Cognitive Skills

General mental skills involving reasoning, problem-solving, planning, and experience-based learning, shown in Figure 2.2, are referred to as cognitive abilities or skills.

https://doi.org/10.1515/9783111191492-002

Cognitive abilities are crucial for problem-solving on the job and for learning new material. Cognitive abilities can be increased over time and grown across a career. They can be described as the processes by which the human brain learns, remembers, analyzes, and solves issues.

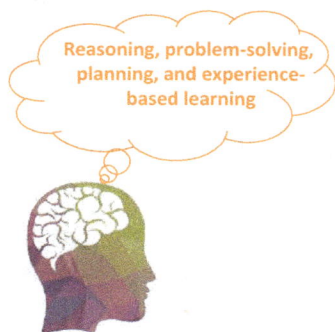

Figure 2.2: Knowledge or cognitive skills.

Cognitive abilities support a variety of tasks at work, including data interpretation, team goal retention, and meeting attendance. In order to remember prior knowledge that might be relevant to the objectives of your organization and to draw significant connections between previously learned material and new information, you need these talents.

When it comes to cognitive abilities, the chemical technician must be able to comprehend and follow technical directions as well as verbal and written chemistry terminology. They must understand how to gather, preserve, and get representative samples as well as how to make analysis plans. They should be able to recognize the proper chemicals, glassware, and equipment needed to conduct the analysis.

2.2 Chemical Laboratory Technique Skills

Chemical technicians who operate in various types of laboratories must be proficient with the use of all balances and analytical equipment. They need to know how to put together lab glassware so they may do experiments like reflux, extraction, and distillation. They occasionally have to use pressurized gas cylinders, regulators, fittings, and tool carts properly. They must possess practical laboratory skills such as weighing and measuring chemicals, purifying materials, running columns, conducting distillations, operating instruments, adhering to safe laboratory procedures, handling and disposing of chemicals, and record keeping. Any chemical technician working in chemistry-based laboratories needs to have six key laboratory skills. Figure 2.3 provides a summary of them. These six laboratory skills are described further.

2.2.1 Laboratory Safety Practices

These are safety practices for the prevention of accidents. It is crucial to know how to properly dispose of chemical waste, respond in an emergency, and identify common hazard symbols.

2.2.2 Analytical Methods

These are analytical techniques that are used to recognize, describe, and measure various chemicals.

2.2.3 Sample Preparation

Sample preparation is the process of taking a portion of a larger whole for analysis. A portion of the sample can be examined for less money and in less time than the entire thing.

2.2.4 Test Methods

Chemistry tests can be performed in a variety of ways, from litmus tests to drug tests. People conduct tests for a variety of reasons as well.

2.2.5 Understand Laboratory Equipment

Pipettes, columns, balances, hotplates, and Bunsen burners are necessary chemical equipment in addition to more expensive instruments like nuclear magnetic resonance spectrometers, Fourier-transform infrared spectrometers, or mass spectrometers.

2.2.6 Documentation and Reporting

Documentation and reporting include gathering information and conducting research to disseminate others or maintain records.

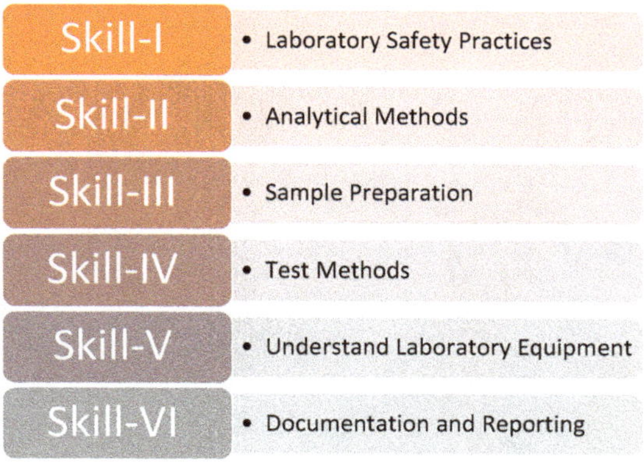

Skill-I	• Laboratory Safety Practices
Skill-II	• Analytical Methods
Skill-III	• Sample Preparation
Skill-IV	• Test Methods
Skill-V	• Understand Laboratory Equipment
Skill-VI	• Documentation and Reporting

Figure 2.3: Six key laboratory skills.

2.3 Synthesis, Characterization, and Preparation Skills

A skilled chemical technician should be familiar with the general principles of synthesis in addition to how to prepare different kinds of molecules. They ought to be capable of making samples and reagents. Additionally, they should practice making solutions, dilutions, purification by recrystallization or chromatographic methods, extractions, buffer preparations, chemical handling safely, and waste disposal techniques.

2.4 Analysis and Measurement Skills

Any chemical technician would benefit from having further knowledge in sample analysis, compound weighing, and solutions measurement. Figure 2.4 summarizes and presents some of these abilities.

2.5 Laboratory and Personal Safety Skills

Chemical technicians must understand how to use all personal laboratory safety equipment, be familiar with and adhere to MSDS regulations, handle glassware, laboratory hardware, and equipment correctly, and follow all disposal rules for chemicals and hazardous waste. The chemical technicians must take the following actions to fulfill the requirements:

– Observe every instruction exactly. When you see the word CAUTION, exercise extra caution.
– Learn where each piece of safety equipment is located in your laboratory. This might include showers, eyewash stations, and fire extinguishers.

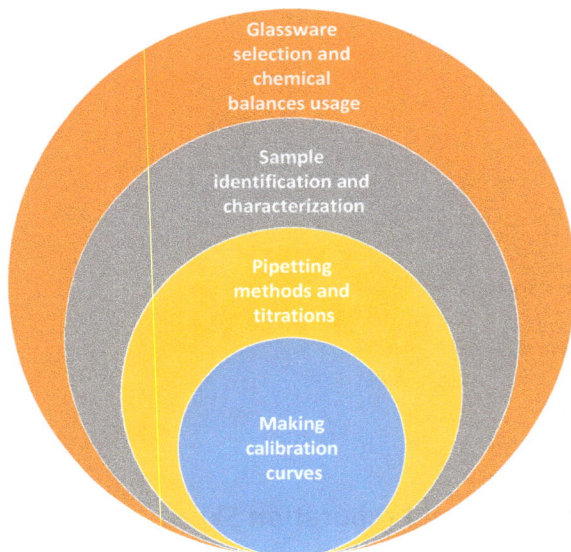

Figure 2.4: Analysis and measurement skills.

- Never run, push, or engage in horseplay.
- Any accidents or safety risks should be reported right away.
- Never conduct unlicensed experiments. In a lab, you should never work alone.
- As directed, dispose of used chemicals.
- Keep work areas tidy and free of any extraneous items.
- Dress appropriately for the laboratory.
- When working with hot or dangerous materials, wear safety goggles.
- When interacting with heat or fire, exercise particular caution.
- Use care working with chemicals and laboratory equipment.

2.6 Housekeeping Skills

These skills include keeping all parts of the laboratory clear of clutter, waste, unnecessary equipment, and unused chemical containers, which is referred to as housekeeping because it pertains to the overall state and look of a laboratory.

For maintaining proper housekeeping in laboratories and typically in any workplace, there are various techniques. The 5S housekeeping method, which emphasizes waste reduction through workplace organization, is one of these techniques. As shown in Figure 2.5, the acronym 5S was created from the Japanese terms "seiri," "seiton," "seiso," "seiketsu," and "Shitsuke."

Figure 2.5: 5S housekeeping method.

The 5S method is a Japanese management approach derived from the Toyota Production System (TPS). Its name is formed from the first letter of each of the five processes. It is founded on five basic ideas that are explained in detail in Figure 2.6.

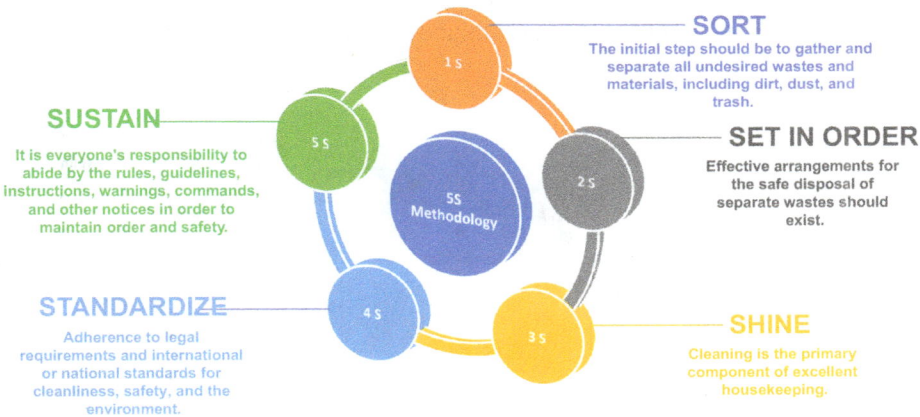

SORT

The initial step should be to gather and separate all undesired wastes and materials, including dirt, dust, and trash.

SUSTAIN

It is everyone's responsibility to abide by the rules, guidelines, instructions, warnings, commands, and other notices in order to maintain order and safety.

SET IN ORDER

Effective arrangements for the safe disposal of separate wastes should exist.

STANDARDIZE

Adherence to legal requirements and international or national standards for cleanliness, safety, and the environment.

SHINE

Cleaning is the primary component of excellent housekeeping.

Figure 2.6: A simple explanation of the five core principles of 5S housekeeping.

The 5S principles are especially successful since they enhance workplace cleanliness and result in the following advantages and benefits:

- Assure increased workplace safety
- Increase the equipment's accessibility
- Reduce the rates of defects
- Reduce spending
- Lower rates of injuries
- Improve employee morale and productivity

The 7S principles are an extension of the 5S excellent housekeeping system. Sort, set in order or systematize, shine or sweep, standardize, safety, self-discipline, and sustain are all abbreviations for the 7S method. Figure 2.7 depicts the seven S's, which are described as follows:

- **Sort**: Involves locating and getting rid of extraneous stuff and unnecessary items in the workplace.
- **Set in order or systematize**: Putting things in a proper arrangement before use.
- **Shine or sweep**: Leaning your workplace and equipment. In order to accomplish this, hard and soft floor finishes can be vacuumed and mopped, laboratory benches can be damp-wiped, waste bins can be cleaned and emptied, and waste bags can be replaced.
- **Standardize**: Setting and upholding high standards for good housekeeping. Keeping an inventory, ordering chemicals and supplies, and performing regular maintenance on and calibrating laboratory equipment are all ways to achieve this.
- **Safety**: Maintaining a secure workplace. In order to do this, choose storage containers that are suitable for the contents to be stored, and use cleaners, acids, and solvents safely.
- **Self-discipline**: Ability to act impulsively without being directed or given orders.
- **Sustain**: Carrying out 7S actions continuously to get consistently good results.

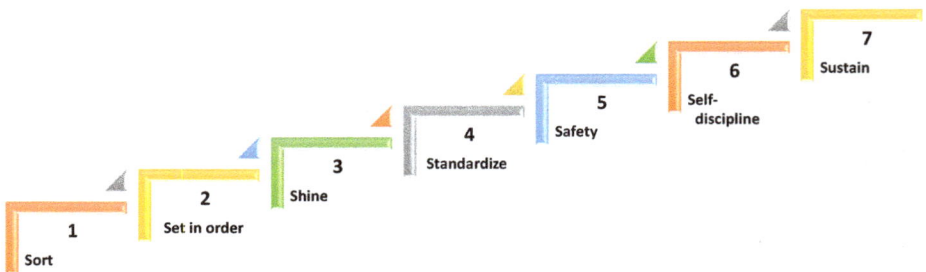

Figure 2.7: The housekeeping 7S principles.

In addition to the 5S and 7S concepts that were previously mentioned, the next 11 tips or suggestions are absolutely essential for efficient workplace housekeeping:

- **Prevent trips, falls, and tripping**: The second most common cause of nonfatal occupational injuries or illnesses requiring days away from work was slips, trips, and falls.
- **Get rid of fire risks**: Employees are in charge of preventing unnecessary accumulations of combustible materials in the workspace. Combustible garbage needs to be "daily disposed of and stored in covered metal containers."
- **Manage dust**: A major explosive hazard exists when dust collection measures more than 1/32 of an inch, or 0.8 mm, and covers at least 5% of the surface of any given place.
- **Do not trace materials**: Work-area mats, which can be made of cotton or have a sticky top, should be kept tidy and in good condition. This helps stop the transfer of dangerous items to other workspaces.
- **Avoid having objects fall**: Toe boards, toe rails, and nets, for example, can help stop objects from falling and striking employees or equipment.
- **Clear the clutter**: Because clutter restricts movement in a crowded workplace, ergonomic issues could develop, and workers could experience injuries.
- **Store items properly**: Storage facilities should not accumulate materials that could cause tripping risks, fires, explosions, or bug infestations.
- **Make use of and examine tools and personal protective equipment (PPE)**: When doing housekeeping, it is advised to wear minimal PPE, such as closed-toe shoes and safety glasses. Depending on the dangers, decide which PPE to wear.
- **Establish the frequency**: Every employee should help out with housekeeping, especially when it comes to keeping their individual workspaces organized, reporting any safety risks, and wiping up spills when necessary. Workers should check, tidy, and get rid of any unnecessary items from their workspaces before their shift ends. According to experts, this dedication can save on future cleaning time.
- **Produce regulations in writing**: The consensus among experts is that cleaning policies ought to be documented. Which cleansers, tools, and techniques to be utilized could be laid out in written procedures.
- **Take a long view**: Housekeeping should be ongoing through monitoring and auditing rather than being a one-time project. To support housekeeping, retain records, follow a regular walkthrough inspection schedule, flag hazards, and teach staff.

2.7 Computational Skills

The ability to do essential addition, subtraction, multiplication, and division problems rapidly and correctly using mental calculations, paper and pencil, and other aids is known as a computational skill. Computational abilities include, but are not limited to:
- the capacity to do basic mathematical procedures;
- knowledge of the dimensions analysis technique, as well as familiarity with the Laboratory Information Management System;
- understanding of how to calculate and convey chemical concentrations;

- the ability to analyze and plot graphic data;
- the ability to locate pertinent laws and standards for resolving issues;
- the capacity to carry out an experiment, assess the outcomes, and reach reliable conclusions;
- the capacity to generate and convey scientific information;
- the capacity to capture experimental data consistently and dependably;
- the ability to learn through appropriate books, materials based on research, or other learning tools;
- in-depth understanding of how computers respond to commands from humans;
- knowing how to use the fundamentals of computer programming to solve a number of problems;
- using the appropriate mathematical techniques to address physics and math challenges; and
- the capacity to use computational methods for the application of theory in practice.

2.8 Computer Skills

These are skills that include comprehension of computer language, functional processes, and concepts. Computer literacy can range from fundamental abilities like turning on and off a computer and utilizing simple email or word processing programs to more complex programming and software development skills. Figure 2.8 provides a summary of some of these abilities.

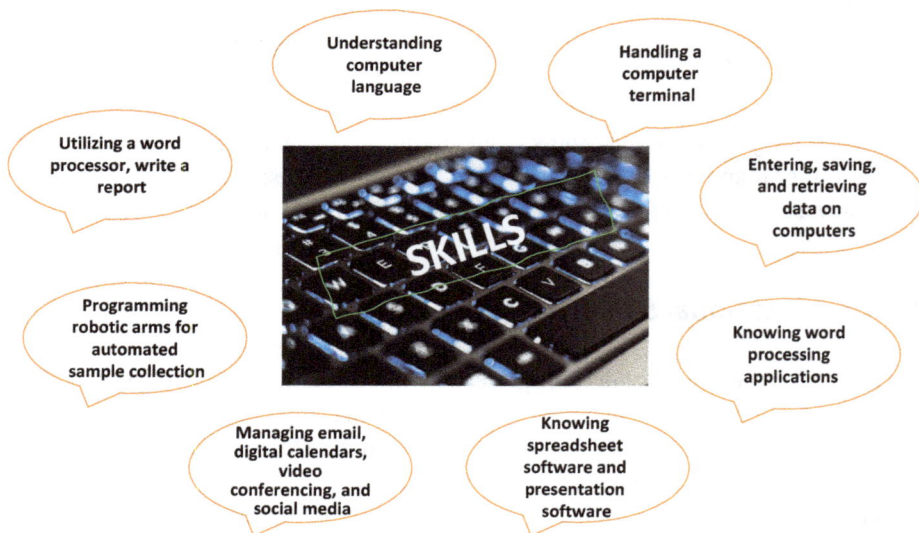

Figure 2.8: Computer skills.

2.9 Communication Skills

A facility cannot run safely or productively without one crucial component, and that crucial element is effective communication. Chemical laboratory technicians communicate with one another, with operators and maintenance workers, with management, and with other people. All parties must comprehend and completely understand all messages. Misunderstandings could lead to both material damage and physical harm.

It is vital to make sure the other person understands what is said while conveying information to them. It is important to understand the information you are given, or else you should keep asking questions until you do.

If the work is done in shifts, the person in charge of the preceding shift will provide information on process operations, reaction conditions, and so on to the incoming plant operation technician or laboratory technician during a shift change or turnover. The new employee should be informed of any modifications since his/her last shift as well as the current state of each system for which he/she is accountable. He/she ought to inquire about any issues that might have arisen during earlier shifts. The new technician needs to be aware of any planned maintenance, other work on the lab, or other work on the equipment in the area for which he/she is responsible. The prior shift technician can depart after the new individual has the information he/she requires. The following is a list of essential practices chemical technicians need in order to have successful communication abilities:

- Maintain the lab's inventory and sample catalog in a proper manner
- Observe things and foresee wants
- Maintain precise recordkeeping
- Establish objectives and carry out lab tasks in collaboration with team members
- Explain the requirement for chemicals, tools, books, and other resources
- Create conclusions for the written report's ultimate product
- Write up and deliver oral reports
- Make written notes in logbooks and record instrument readings on data sheets
- Keep records of instrument maintenance and calibration

2.10 Questions

2.10.1 What are the possible chemical technicians' workplaces?
2.10.2 Why communication skills are crucial?

2.10.3   List down five abilities laboratory technicians require as part of computer skills.

2.10.4   What are the housekeeping 7S principles?

2.10.5   Where from the 5S housekeeping method was created?

2.10.6   What are the benefits of the five core principles of 5S housekeeping?

2.10.7   What are the six key laboratory skills?

2.10.8   Give three examples for cognitive skills.

2.10.9   What is the role of the chemical laboratory technician in research and development?

2.10.10 List down five essential tips for efficient workplace housekeeping.

Chapter 3
Chemical Plant Operation Technician Skills

Chemical plant operation technicians, as excellent communicators and service providers, must have other skills such as problem-solving, preventative maintenance knowledge, hand tool usage, and safety regulation awareness in addition to the following core skills:

3.1 Knowledge Skills

One of the most crucial workers in the plant is the chemical plant operation technician. He must be aware of how process systems work and have the ability to identify any issues with process machinery. He/she needs to keep an eye on the plant's machinery and equipment in order to spot any issues that could arise when the processing system is in operation. He/she must also be able to change process variables like pressure, temperature, level, and flow as needed, which may be possible in some workplaces, as well as put the system into a safe mode in the event of a malfunction.

3.2 Technical Skills

The technicians who manage chemical facilities typically do a number of production tasks, such as preventing and fixing problems as they occur. If a technician notices that a pump bearing is about to break and interfere with production, he or she could swap it for a backup pump.

Chemical technicians can also practice preventative maintenance to help avert problems. Preventative maintenance is carried out to stop equipment breakdown. Standard preventive maintenance practices include cleaning or changing equipment filters, adjusting packing on valves, and lubricating machinery.

3.3 Personal Safety Skills

The ability to create and implement behavior-based safety programs should be a competency for chemical technicians who operate as safety technicians. To preserve secure working circumstances, this entails developing a set of guidelines that staff members must adhere to.

For instance, if an employee has an injury at work, he/she can be rewarded for adhering to their treatment plan and reporting to work earlier than expected. Technicians

https://doi.org/10.1515/9783111191492-003

can also track workplace incidents and pinpoint areas that require improvement by using behavior-based safety initiatives.

Technicians are required to apply a variety of regulations to the disposal of both hazardous and nonhazardous waste as part of their personal safety duties, and they are always required to do so. Regulated garbage that poses a risk to both people and the environment is known as hazardous waste. Technicians should constantly try to produce the least amount of garbage possible, both hazardous and nonhazardous waste. For instance, technicians should only take what is necessary when sampling. Taking too much material is wasteful.

Technicians should also be on the lookout for potential problems that could result in waste. For example, if a technician notices a corroded drum, he or she should report it before it leaks and causes a larger problem. Technicians should ensure that any materials left over from a process are properly routed. Special sumps, drums, and disposal tanks are used in many plants to remove and store hazardous materials. Separation of hazardous and nonhazardous materials is required. If hazardous waste is mixed with nonhazardous waste, the entire mixture must be treated as hazardous waste from that point forward.

3.4 Collaboration Skills

Teamwork is the process of working with a group of people to complete a task within a business, ensuring that employees and management work together by maximizing their talents and providing constructive feedback. Chemical technicians must be effective and considerate team members in order to increase output, morale, and a happy workforce. They must act in an environment of mutual trust and cooperation while employing the appropriate strategies. A few of these abilities are described in greater detail and illustrated in Figure 3.1.

- **Long-term perspective**: The essence of collaboration may be summed up as focusing on a common objective while assessing how each individual contribution fits into the overall scheme. Vision and long-term thinking are crucial because of this.
- **Broad-mindedness or open-mindedness**: Accepting new ideas and being open to suggestions is the ultimate foundation for effective collaboration. Each person can contribute something valuable, and many come from cultural and social backgrounds that are so fundamentally different that they affect the very structure of the team.
- **Communication**: Straight and unambiguous communication is essential for productive teamwork. A team, by definition, is a collection of individuals who must work closely together and maintain open lines of communication. People converse differently, while some are comfortable speaking in front of groups and engaging

Figure 3.1: Six critical collaboration abilities.

in discussion, others are more reserved. Fostering effective communication results in the best work atmosphere and, ultimately, the best results for a team's work.

– **Organization**: Only when people assign jobs, divide workloads, take care of their duties, and plan their activities, collaboration can be successful. As a result, being organized is a crucial skill for collaboration.

– **Flexibility or adaptability**: Life is multifaceted and highly complex, so nothing always goes as expected. This regulation applies to project collaboration as well. Priorities change challenges block development, and a number of issues may completely derail a project. Therefore, flexibility is a valuable quality for teamwork.

– **Deliberation or debate**: Effective team collaboration requires meaningful and fruitful in-team dialogue. To start afresh, you must be able to compromise when necessary and refrain from becoming overly committed to your beliefs. Discussion often leads to creation because it enables everyone to evaluate and expand upon good ideas while discarding bad ones.

The following six methods can be used to develop and improve these collaborative skills and talents in the workplace:

– Goals should be clearly specified for each project, phase, financial year, or time period. Also, duties should be clear. New hires must be aware of their obligations. It is important to ensure that clear limits have been established to specify work outputs and responsibilities in order to keep employees motivated.

–   Employees can be assisted in developing their strengths by utilizing and showcasing each person's skills while distributing work. Giving them a sense of worth inside the group will motivate them to offer their special talents to the common objective. Finding out how each person feels and where he/she can contribute more effectively takes a lot of work.
–   The success of the team is influenced by providing employees with learning opportunities, allowing them to advance in their careers, and giving them space to pick up new skills that will benefit them personally.
–   By ensuring that all channels of communication are available, whether through in-person interactions or online forums, an environment of openness can be promoted. Staff members are encouraged to participate in discussions, contribute ideas and opinions, and abandon the "no stupid questions" policy that severely curtails innovation in this setting.
–   Transparency, trust, and encouragement are the cornerstones of effective teamwork. Building trust is a sure-fire method to be transparent. The finest methods for overcoming hurdles and potential impediments are honesty and transparency.
–   Employee concentration is made simpler when teams are provided with the proper hardware and software, such as document automation tools, and role-specific tools.

3.5 Documentation Skills

The ability to maintain accurate records of their work and the safety precautions they have taken is a must for safety technicians. For each piece of equipment in a workplace, this can entail keeping track of inspections, maintenance schedules, and training materials. You can detect any potential dangers or hazards at your job site and take action to mitigate them by keeping thorough records.

The laboratory technician could require a particular skill set if his/her job involves creating and disseminating documents. Documentation duties can be finished more quickly by using skills like reporting, presentation, and organizing. Employers can be impressed by being more knowledgeable about these abilities and how to develop them. Technicians who are skilled in documentation can provide clear, polished documents. Figure 3.2 gives a brief overview of certain traits and abilities that might cultivate to improve documentation.

For better writing and proper understanding of documents, consider the following suggestions as you exhibit your documentation skills at your workplace:
–   Create visually appealing documents. Although the content of your documents should be the main focus, making sure they are visually appealing will help you grab readers' attention. Also, it might help make your article straightforward to read, allowing readers to quickly scan it wherever feasible to save time.

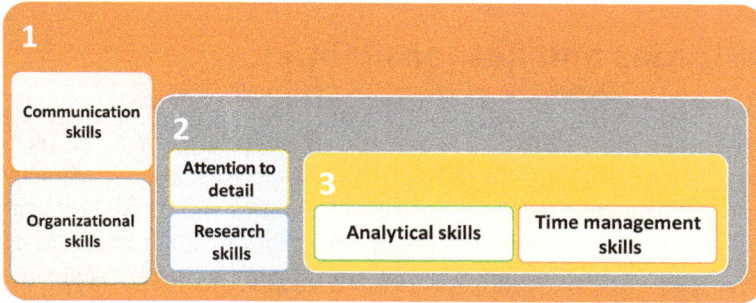

Figure 3.2: Skills and attributes help in improving documentation.

– Limit your use of technical terms. When creating documents for a general audience, try to limit the usage of technical jargon and acronyms. This ensures that your publications will be understandable to all viewers, and you can also add pictures to your documents to make them look better.

3.6 Questions

3.6.1   What are the six critical collaboration abilities?
3.6.2   Define teamwork.
3.6.3   Give examples for personal safety skills.
3.6.4   List down three chemical plant operation technician core skills.
3.6.5   What does open-mindedness mean?
3.6.6   List down two methods that can be used to develop and improve collaborative skills and talents in the workplace.
3.6.7   Give three traits and abilities that might cultivate to improve documentation skill.
3.6.8   Explain how the visually appealing documents help at workplace.
3.6.9   How limiting the use of technical terms help in understanding documents?
3.6.10 What are the cornerstones of effective teamwork?

Chapter 4
Technician Duties and Responsibilities

The foundation of a scientific research lab is based on its laboratory personnel. Technicians may operate alone or as a member of a scientific staff team, and the majority of their work is performed in laboratories. The tasks a laboratory technician performs are mostly determined by the region in which they work. They might examine crude oil samples or any other relevant samples if they work for a gas and oil company. They might perform blood tests, analyze body tissues or fluids, and look at cells if they operate in a medical setting. If they are employed by a company that produces food and beverages, they may test food samples to look for contamination or guarantee quality. Environmental agencies, specialized research organizations or consultancies, universities, hospitals and clinics, the water and pharmaceutical industries, and many more are among the many employers of laboratory technicians.

A laboratory technician's general duties include:
- running and supporting scientific research and experiments;
- designing, implementing, and carrying out controlled experiments;
- writing reports, reviews, and summaries;
- collecting, preparing, and testing samples;
- maintaining, calibrating, cleaning, and testing the sterility of the equipment;
- providing technical assistance, and being current with pertinent scientific and technical advancements; and
- purchasing and keeping track of supplies and resources.

4.1 Chemical Laboratory Technician

Working in a lab, a chemical lab technician helps chemists and chemical engineers with various kinds of activities. The tasks of a lab technician include planning, supervising, and conducting tests as well as writing a report on the findings. These tests are frequently carried out to investigate the qualities or security of novel chemical goods. Strong analytical abilities, familiarity with common laboratory tools, and the capacity for teamwork are the main requirements for a career as a chemical lab technician.

4.1.1 Duties and Responsibilities

Chemical lab technician duties and responsibilities include:
- conducting laboratory research in accordance with standard practices;
- recognizing and fixing equipment problems;

https://doi.org/10.1515/9783111191492-004

- keeping all tools and equipment used in laboratories clean, secure, and in good working order;
- producing standardized reports on laboratory experiments;
- executing lab tests in accordance with equipment requirements and standards;
- placing orders for lab equipment and supplies;
- creating novel laboratory techniques;
- adhering to correct garbage disposal and recycling practices; and
- creating reports on test results and writing experimental protocols.

4.2 Environmental Laboratory Technician

Environmental lab technicians run tests in the lab to monitor and identify contaminants. Professional lab testers examine the environment's health as well as the health of people and animals. Outside of the lab, some lab staff members might collect samples.

Environmental laboratory technicians can have positions as environmental specialists, laboratory specialists, environmental technicians, and environmental health professionals. Environmental laboratory technicians monitor both indoor and outdoor environments in cooperation with scientists and engineers to prevent environmental dangers or contaminants.

They are in charge of collecting samples of soil, air, and water for laboratory analysis. On rare occasions, they may be tasked with checking potential construction sites to ensure they are tidy before work begins. They also assess any potential environmental impacts of the construction.

Most technicians are employed by environmental testing facilities, advising firms, or regional or municipal governments. Those who specialize in lab work spend most of their time indoors in labs and offices, whereas those who work in field collection spend most of their time outdoors in a range of locales and weather conditions, investigating everything from urban industry sites to remote rivers and lakes. Fieldwork may require a significant amount of walking, standing, carrying, and lifting. Environmental laboratory technicians frequently have full-time jobs, and they may occasionally need to work irregular or extended hours in the field.

Although the majority of jobs for environmental laboratory technologists are with governmental organizations that provide organizations trying to comply with environmental regulations with expert advice, environmental lab technicians can also obtain employment with private businesses. Figure 4.1 provides an overview of some of the most common sectors requiring environmental laboratory technicians.

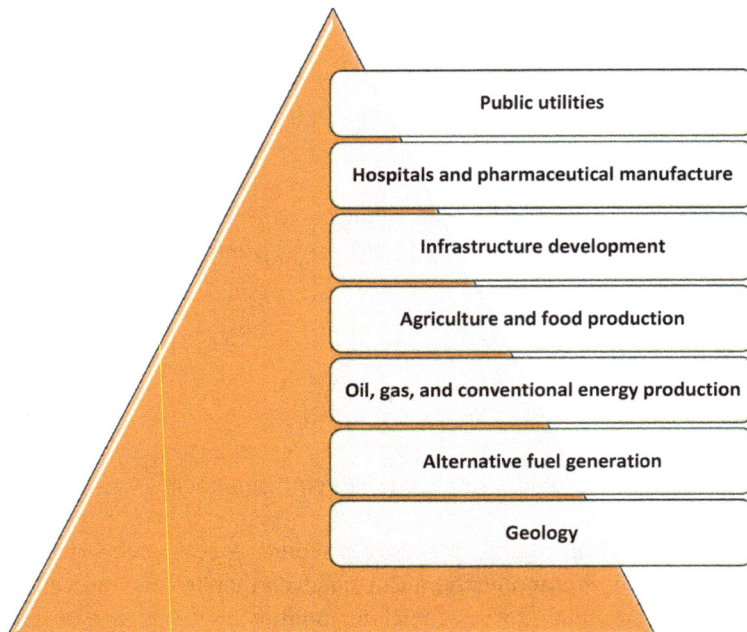

Figure 4.1: Most popular industries for environmental laboratory technicians.

4.2.1 Duties and Responsibilities

Duties and responsibilities of environmental laboratory technicians include:
- testing drinking water, the ground, and the surface to see whether any toxins are present;
- conducting tests that look for both organic and inorganic substances and providing associated assessments for air and soil;
- taking samples from commercial or industrial locations, public lands, or nearby places;
- setting up monitoring devices that gather essential data;
- using said monitoring devices;
- providing data entry services;
- conducting safety audits of your own or your client's facilities;
- putting in place OSHA-compliant safety measures, and sterilizing lab apparatus in accordance with industry requirements;
- keeping sufficient inventories of technical supplies and tools; and
- aiding with technical matters to experts working for outside groups.

4.3 Food Science Laboratory Technician

Technicians in food science laboratories, often known as food technologists, do research to advance methods for the safe manufacture, storage, and transportation of food products. Their principal employers are food firms, though they may also work for the government, businesses that supply the food industry, or businesses that produce food equipment. Their major responsibility is to create novel food or flavor varieties. To do this, they produce food prototypes that will be evaluated by a human panel. In addition, they assess the product's price, shelf life, and manufacturing ability. Additionally, they assist with organizing and carrying out the extraction and purification process. They keep an eye on and ensure that the laboratory's manufacturing process is effective, efficient, and clean. Additionally, the laboratory technician may perform standard duties including preparation, measurement, and packing of extract commodities.

4.3.1 Duties and Responsibilities

Food science laboratory technicians must perform a variety of tasks. They identify foods, keep track of them, and report on their composition and quality. Food scientists and technologists can also assist in the improvement of the taste and texture of a new or existing food product, the planning of food production techniques, and the conduct of exploratory research and experiments on foods and beverages. They are capable of performing the following additional tasks:

- examine the color, consistency, fat content, caloric content, and nutrients of the food;
- examine test results and compare them with standard caloric and nutritional content tables;
- examine samples to identify cell structures or bacteria or foreign material;
- make reagents, combine, boil, cut, blend, separate, or freeze ingredients;
- maintain, clean, operate, and sanitize laboratory equipment such as microscopes and Petri dishes;
- determine the amount of moisture, salt, sugar, or preservatives in food and beverages;
- use mathematical and chemical procedures to calculate ingredient and formula percentages;
- keep track of or compile test results, as well as create graphs, charts, and reports;
- inspect food, food additives, and food containers for compliance with established safety standards;
- help with food research, development, and quality assurance;
- ensure that food products are suitable for distribution;

- contribute to the development of proper food packaging, including bottles and plastics;
- examine foods, chemicals, and additives to see if they are safe and have the right combination of ingredients;
- understand and follow procedures, methodologies, and compliance steps in the ingredient development process;
- extract plant material using standard operating procedures (SOPs) that have been developed;
- oversee large machinery used in raw material preparation, purification, and extraction;
- ensure product development consistency and efficacy by adhering to SOPs and maintaining accurate documentation of work completed;
- assist in the production of batch records and labels;
- assist with the weighing, packing, labeling, and documentation of all development ingredients, as well as the tracking of batch numbers, lot numbers, and expiration dates.
- Comply with all health and safety, Good Laboratory Practices, and Good Manufacturing Practice guidelines.

4.4 Wastewater Treatment Laboratory Technician

A wastewater treatment laboratory technician conducts laboratory tests on raw and treated wastewater and documents the results and data. He/she tests wastewater samples at various stages of treatment on a daily basis to ensure compliance with environmental standards. He/she is also in charge of operating, cleaning, calibrating, and maintaining laboratory equipment and instruments. He/she calculates results from raw laboratory data, logs results, and keeps accurate test records to help with the preparation of various wastewater treatment reports. As needed, he/she prepares reagents, standards, and test solutions. Additionally, he/she resolves day-to-day problems from treatment plant operators regarding appropriate materials, methods, and procedures for collecting treatment plant samples.

This technician must learn about laboratory equipment and instruments, as well as how to use them safely and effectively, chemical hazards and safety precautions, testing potentially hazardous substances, chemistry and biochemistry, and Microsoft Office suite. He/she must also establish and maintain effective working relationships with other staff and the general public, communicate effectively both orally and in writing, keep accurate records, prepare a variety of reports, understand and use chemical and mathematical equations and calculations, follow complex oral and written instructions, procedures, and guidelines, and operate and maintain standard laboratory equipment, automatic samplers, and instruments.

4.4.1 Duties and Responsibilities

Aside from the tasks mentioned above, wastewater treatment laboratory technicians must also complete the following:

- Collect and store wastewater samples for required analyses; analyze for contamination using a variety of EPA-approved methods and procedures; and interpret test results.
- Maintain a record of test information, results, and procedures.
- Ensure that the plant is following local treatment regulations; notify the supervisor of any procedures, analysis results, or trends that could result in a regulatory violation.
- Perform preventative, routine, and servicing operations for the city's wastewater treatment plant.
- Perform specialized laboratory tests for plant operations and outside agencies.
- Collect and test wastewater samples as needed for plant efficiency reports.
- Respond to emergencies involving wastewater system failures, leaks, or other issues; assess the situation and make necessary repairs.
- Modify or repair instrumentation and control equipment, such as recorders, flowmeters, and other water quality monitoring equipment.
- Operate and inspect equipment, instruments, pumps, valves, meters, and other apparatuses; control and adjust flow and treatment process using pumps and valves.
- Assist in the daily operation, maintenance, and repair of a wastewater treatment plant; collect water samples and perform lab analyses.
- Be responsible for routine custodial and ground maintenance duties at the wastewater treatment plant.

4.5 Pharmaceutical Laboratory Technician

A pharmaceutical lab technician works in a laboratory, assisting those looking for the next miracle drug and doing the grunt work required to discover the perfect cure. Despite the fact that most people only hear about the finished product, developing a new drug can be a time-consuming and laborious process. A pharmaceutical lab technician is in charge of collecting samples, creating slides, and organizing experiments. He/she also weighs samples, runs tests, collects results, and accurately records them.

A pharmaceutical lab technician is in charge of a variety of tasks in the pharmaceutical industry. He/she frequently collaborates with scientists to ensure that drugs and other products meet the quality standards before they are made available to the general public.

A pharmaceutical lab technician may also be assigned to test new drug formulas or create prototypes for new products. In addition to his/her hands-on duties, he/she

frequently spends time performing administrative tasks such as equipment maintenance and inventory tracking.

4.5.1 Duties and Responsibilities

In addition to the duties listed above, pharmaceutical lab technicians may be responsible for the following:
– carrying out routine laboratory tests to ensure that products meet the quality
  standards;
– keeping track of test results and identifying issues or inconsistencies with samples or procedures;
– cleaning equipment and maintaining supplies to keep the lab in good working
  order;
– keeping a supply inventory and reordering items as needed;
– assisting with the setup and takedown of equipment as needed;
– preparing samples for analysis by collecting, measuring, mixing, titrating, filter-
  ing, weighing, or labeling chemicals;
– mixing solutions and preparing reagents for studies as directed by scientists and
  lab managers;
– verifying the correctness and consistency of data charts and reports produced by
  instruments like spectrophotometers and chromatographs;
– adhering to safety protocols when handling chemicals or equipment to lower the
  chance of mishaps or injuries; and
– place orders for, label, and count supplies or chemical stock, then input inventory
  data into a computer.

4.6 Medical Laboratory Technician

Medical laboratory technicians support doctors and other healthcare professionals.
They use laboratory tools and computers to examine physiological fluid samples,
human tissue samples, and abnormalities in order to identify and classify them. Medical technicians and other similar professionals who work in the healthcare industry
are trained in medical sample collecting and laboratory testing. Their work is crucial
to the patient's care. Education, responsibilities, and certifications attained define a
medical laboratory job title.

4.6.1 Duties and Responsibilities

Medical lab technicians are individuals who carry out standard medical laboratory procedures, providing doctors with the knowledge they need to identify, treat, and prevent disease. They can ascertain the chemical composition of body fluids, search for parasites, bacteria, and other microbes, search for aberrant cells in the blood and other body fluids, match blood for transfusions, and test for medication levels in the blood by looking at and analyzing body fluids and cells.

They make use of cell counters, microscopes, and other high-tech lab apparatuses. Additionally, they make use of automated and computerized tools that may run several tests at once. They examine the findings after testing and inspecting a specimen, and then they present those findings to the doctor. They constantly set up, keep up, calibrate, clean, and test the sterility of laboratory equipment. Additionally, they use automatic analyzers to conduct chemical analyses of bodily fluids like blood or urine in order to look for abnormalities or diseases. They then enter their findings into a computer and analyze the data to make sure that the tests or experiments they conduct conform to the specifications.

Medical lab technicians work in physician offices, hospitals, and private labs and perform standard clinical laboratory tests in hematology, chemistry, immunohematology, microbiology, immunology, and coagulation at the beginning of their career. Analysis levels can range from simple point-of-care testing to comprehensive testing that involves all significant clinical laboratory domains. They also perform a variety of preanalytical, analytical, and postanalytical process-related tasks. They also have additional duties in the laboratory's quality control, training, and information processing departments.

They should possess strong communication abilities required for consultative interactions with other members of the healthcare team, technical service matters, and patient education. They should also exhibit the moral character required to protect patient privacy, and win over the trust of clients, business partners, and the community. Other duties typically performed by medical laboratory technicians include the following:
- using established formulas or experimental protocols, prepare standard volumetric solutions or reagents to be mixed with samples;
- instruct or supervise other technicians or laboratory assistants;
- get tissue or blood samples from patients while adhering to the asepsis guidelines;
- evaluate and document test results to provide reports with graphs and charts;
- test raw materials, processes, or finished products to determine the nature, quantity, or qualities of a substance;
- follow several processes to collect samples, culture, isolate, and identify microorganisms for examination;
- examine stained cells using dye to check for irregularities;
- do blood tests related to transfusions and blood counts;

- conduct medical research to treat or prevent disease;
- engage in routine sample management, collection upkeep, and instrument and quality assurance; and
- gather and analyze data, maintain detailed records, report data, and comprehend fundamental concepts.

4.7 Science Laboratory Technician

A science laboratory technician is a person who works in laboratories to assist scientists. Chemical, biological, agricultural, environmental prevention, forensic, forest and conservation, geological, and energy technology are some of the science laboratory technology fields. This technician frequently entails working with complex systems to aid in the operation of scientific processes and projects, to properly record the results, and to assist in the routine procedures that take place in a laboratory. He/she also assists scientists with their investigations by measuring, recording, and analyzing various scientific data. He/she can also provide direct technical support to the teaching and learning of science-based subjects, such as risk assessments, ensuring that all laboratory science-based teaching activities are carried out in accordance with current policies and best practices.

4.7.1 Duties and Responsibilities

The nature of the work will vary depending on the organization; for example, in an environmental health department, you may be involved in analyzing food samples to consider prosecution and protect public health, whereas, in the water industry, your work will primarily focus on water sample collection and analysis. The following are the primary duties and responsibilities:
- assist in chemical analysis in educational institutions' laboratories, food and chemical industries' laboratories, and research institutes;
- assist in biological experiments and investigations in industrial and institutional laboratories, farms, museums, and other nature establishments;
- perform laboratory tests in order to generate accurate and reliable data to support scientific investigations;
- perform routine tasks according to known methodologies in order to conduct analyses;
- prepare specimens and samples;
- standardize and maintain laboratory equipment such as centrifuges, titrators, pipetting machines, and pH meters; and
- maintain clean and functional equipment while ensuring safe waste removal.

4.8 Polymer Laboratory Technician

Polymer technicians work in businesses that manufacture and process polymers like plastic or rubber. Manufacturing, quality assurance, product development, and sales and marketing are all possibilities. Production workers diagnose, analyze, and resolve production-line issues. They may also be in charge of designing and developing polymers, from the time they are created in laboratories and industrial plants until they are used in the manufacturing industry and transformed into the appropriate end products. Their research focuses on composite materials or specific types of other materials such as graphite, metals and alloys, ceramics, plastics and other polymers, natural materials, recycling, and biotechnology applications in biomaterial production.

Polymer technicians' primary goals at work are to perform chemical and physical tests to assist manufacturers in developing and improving polymer properties. Furthermore, they contribute scientific knowledge and methods to the production of a diverse range of products used by consumers, industry, and government. In the workplace, ingredients will be mixed, chemical reactions will be performed, and the polymer's durability and tensile properties, as well as the product's ability to withstand tension and elasticity, will be tested. They also record and share test results with chemists, chemical engineers, mechanical engineers, or other scientists on the research or production team. Figure 4.2 depicts a typical polymer laboratory technician's experiments.

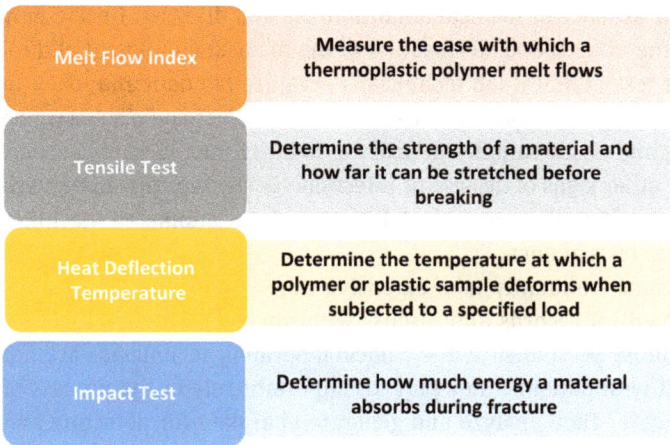

Melt Flow Index	Measure the ease with which a thermoplastic polymer melt flows
Tensile Test	Determine the strength of a material and how far it can be stretched before breaking
Heat Deflection Temperature	Determine the temperature at which a polymer or plastic sample deforms when subjected to a specified load
Impact Test	Determine how much energy a material absorbs during fracture

Figure 4.2: Common polymer material tests.

4.8.1 Duties and Responsibilities

– They make material selection recommendations based on the design brief, such as strength, weight, heat resistance, electrical conductivity, and cost.
– They formulate and synthesize polymers in the laboratory, pilot plant, and industrial scale, and enhance organic and polymer synthesis methods.
– They conduct tests on the new products' physical and chemical properties.
– They provide technical assistance in the development of new products.
– They create documentation for the various tests while keeping the database for these up to date.
– They lead the polymer production tests and process control procedures.
– They resolve any problems that may arise during the chemical process.
– They examine the technical specifications and economic factors associated with the process or product design goals.
– They monitor the materials' quality and assess their deterioration.
– They analyze data pertaining to product flaws and laboratory test results.

4.9 Clinical Laboratory Technician

A clinical laboratory technician is someone who performs medical tests such as blood sampling, urinalysis, and microbiological examinations. These tests are typically used to detect the presence or absence of medical abnormalities and diseases. He/she may also be responsible for the upkeep and sterilization of the medical equipment and instruments used in these tests. Clinical lab technicians organize lab data and use it to assist a physician or pathologist in making a medical diagnosis. He/she also examine and analyze body fluids and tissue samples in order to identify microorganisms, bacteria, abnormal cells, or other signs of disease or infection. He/she typically use a variety of lab equipment to perform the necessary lab tests such as automated machines that can perform multiple tests at once.

Clinical lab technicians frequently take extensive notes about procedures and findings to include in medical records and discuss with physicians. They supervise the work of other laboratory personnel. While clinical laboratory technicians are not usually in direct contact with patients, they play an important role in the process of providing personalized care. They analyze and generate vital data for detecting and treating heart disease, diabetes, cancer, and a variety of other diseases. Some clinical lab technicians specialize in virology, blood banking, histology, and cytology. Most clinical lab technicians in large hospitals specialize in one area. They carry out a variety of tasks in small laboratories. Clinical laboratory technicians work in a variety of settings, as illustrated in Figure 4.3.

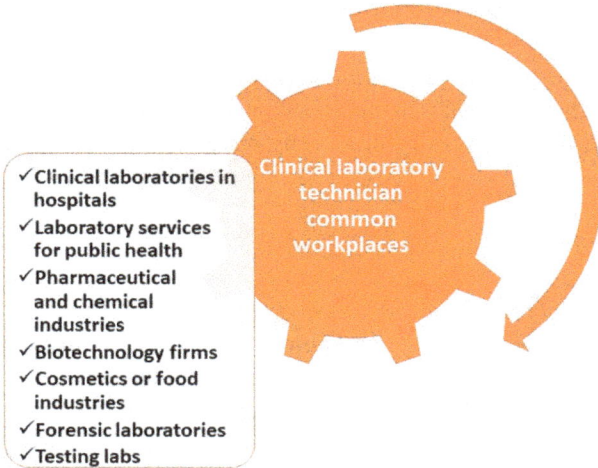

Figure 4.3: Clinical laboratory technician's common workplaces.

4.9.1 Duties and Responsibilities

A clinical laboratory technician's duties and responsibilities are as follows:
– cleaning of glassware and equipment;
– storing equipment and chemicals;
– maintaining and calibrating testing equipment;
– using standard laboratory procedures and formulas to prepare stains, reagents, and solutions;
– preparing, labeling, and storing specimens effectively and analyzing data with manual and automatic equipment;
– following company policies and procedures, as well as applicable laws and regulations;
– keeping test records and communicating results to relevant parties;
– maintaining patient confidentiality while safeguarding data;
– identifying the chemical components of the patient's bodily fluids;
– measuring drug levels in the blood to determine the efficacy of specific treatments;
– assessing the accuracy of test results and interpreting them for the physician;
– entering data from medical tests into a patient's medical record; and
– discussing the findings and results of lab tests and procedures with doctors.

4.10 Quality Assurance and Quality Control Technician

Both technicians are in charge of implementing and approving laboratory procedures, equipment, and processes in order to provide competent and dependable quality control services to labs, plants, and workplaces in a safe and cost-effective manner.

4.10.1 Duties and Responsibilities

–   They lead laboratory equipment reliability and product quality-related improvement tasks, assisting in technical customer complaint investigation, performing root-cause analysis for quality incidents, and participating in the quality meeting.
–   They support the preparation of capital expenditure, operating expenditure budgets, key performance indicators (KPIs), and business plans based on laboratory objectives, and ensure the smooth operation of the laboratory.
–   They assist in the management of laboratory waste.
–   Testing and evaluation are used to introduce and approve new raw materials.
–   They audit the relevant department to ensure raw material quality.
–   They maintain the most recent standard test methods.
–   They review, validate, and develop analytical methods, as well as train staff on test methods.
–   They ensure that SOPs are followed during the execution of the job for laboratory equipment.
–   They prepare and manage an equipment maintenance and calibration schedule.
–   They obtain the necessary standards from suppliers for equipment calibration and validation.
–   They analyze statistical quality control and data, and complete process control reports.
–   They prepare and review the sampling schedule, as well as optimize the sampling requirements.

4.11 Questions

4.11.1   List down three laboratory technician's general duties.
4.11.2   List five duties and responsibilities of the chemical lab technician.
4.11.3   What are the main tasks of the environmental lab technicians?
4.11.4   What are the most common sectors requiring environmental laboratory technicians?
4.11.5   What are the main tasks of the food science laboratory technicians?

4.11.6   What kind of tests do the wastewater treatment laboratory technician conduct?
4.11.7   List down two duties for the pharmaceutical lab technician.
4.11.8   What are the typical polymer laboratory technician's experiments?
4.11.9   What are the polymer technician's primary goals?
4.11.10 What are the clinical laboratory technician's common workplaces?

Chapter 5
Sampling Methods and Chemical Analysis

5.1 Introduction

The term "bulk system" refers to the material under investigation. It is impractical to examine the entire system in the chemical laboratory when analyzing a bulk system. For example, we cannot bring all of the soil for laboratory analysis in order to calculate the phosphate content. As a result, we gather a sample of the bulk system and send it to the laboratory for analysis. As a result, sampling aids in determining the composition of a large amount of analyte by removing a representative for analysis. The goal of sampling is to collect a portion of material that accurately represents the sample's composition. Figure 5.1 shows the six key steps that influence the quality of data: establishing a clear purpose for a sampling program, getting representative samples, handling and storing samples correctly, adhering to the proper chain of custody and sample identification, and accurately analyzing the sample.

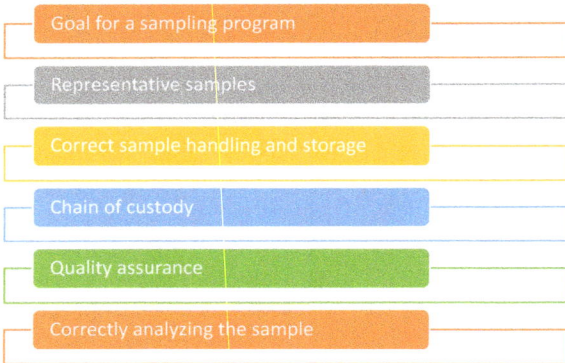

Figure 5.1: Six major activities that determine data quality.

A sample program's design and implementation are significantly influenced by the quantity of the material to be examined, the variety of the ingredients, the lab's accuracy, the cost of the ingredient's analysis, and the ingredient's value. As a result, it is critical to consider the intended purpose of the sampling, the laboratory tests the samples will undergo, and the characteristics of the raw materials and finished goods while developing the sampling operations. Sampling techniques should follow these standards, and laboratory methods should be created and validated in accordance with scientifically accepted principles. Prior understanding of the product data and sampling resources also helps in the selection of the appropriate sampling strategies. The use of recognized international sampling techniques will ensure a unified administrative and

https://doi.org/10.1515/9783111191492-005

technical approach. This will make assessing analysis results on feedlots or consignments simpler.

5.2 Sampling Importance

The benefit of sampling in a chemistry lab is that it allows you to calculate the appropriate response rate based on the total number of points you have selected. It will therefore be utilized in the research study which ought to be sufficient to support extrapolating the findings from the target regions. Sampling has the ability to yield more specific information in addition to being practical and affordable – faster and less expensive. Also, it is more accurate and saves time. It may assist in identifying and understanding cutting-edge marketing frameworks that call for study. Also, it enables scientists to make decisions and draw conclusions from limited data.

5.3 Sample Collection

The most important aspect of any sampling task is to get a sample that accurately represents the bulk system. The sample must meet all requirements for the bulk system in terms of the analyte and analyte concentration. The concentration level for the entire system is assumed to be whatever concentration level is discovered for a specific component of a sample. Obtaining a sample for analysis involves a variety of sampling methods, each with varying degrees of difficulty:
–   depending on the kind of sample that needs to be collected;
–   the homogeneity of the sample's source; and
–   the systems' placement and accessibility.

In most cases, samples can be taken manually and placed straight into a container, especially if automatic samplers are not accessible or if samples are being taken for proper away-use field testing. Another option is to utilize an intermediary vessel. One clean gadget or vessel is lowered into the sample medium using a rope, pole, or chain to collect the sample. In some circumstances, using a hand- or power-operated pump is preferable.

In some instances, automatic samplers are used to collect the samples. There are several different samplers that can be bought commercially. Automatic samplers are more helpful in sampling numerous sites since they minimize human mistakes and keep the samples cool while collected. The components of the sampling equipment should be chosen in accordance with the requirements:
–   Use plastic, glass, Teflon, stainless steel, aluminum, or brass for **inorganic materials** that do not require preservation.
–   Use brass, plastic, glass, Teflon, stainless steel, or aluminum for **nutrition**.

- Use Teflon or stainless-steel plastic for *trace metals.*
- Use plastic, glass, Teflon, stainless steel, aluminum, or brass for *extractable organics.*
- Use glass, Teflon, or stainless steel for *volatile organics.*
- Use a sample container that has already been sterilized for *microbiological samples.*

5.4 Sample Storage

The gathering and evaluation of samples is a crucial part of chemical analysis. The handling and storage of samples have a significant impact on the quality of the results. To obtain high-quality results, the proper storage strategy must consider the sample treatments, temperature, and storage container.

The quality of the sample could degrade if specific factors are not considered when it is collected. If the targeted sample is not kept in a particular kind of container, the results will deteriorate over time. Using a tightly closed vial will keep the sample clean and maintain its integrity. By collecting the sample in an uncontaminated, dry, and suitable container, the risk of diffusion and contamination is decreased. There will be truly little interference from outside sources if the atmosphere is suitable.

The total analysis of the material depends heavily on a quick analysis (within 24–48 h). The accuracy and precision of the results will grow the faster the data is collected and processed. In many cases, the container should be stored in a dark, cold area to prevent any contamination. Further treatments can be added when the environment is insufficient to maintain the sample's quality. The required components, the analytes being examined, and the sort of analysis determine the preservation method. Strong acids must be used if certain metals in a water sample are being analyzed. Nitric acid, for instance, inhibits the oxidation of metal cations. On the other hand, if food samples are taken for analysis, they should be delivered to the lab in the same cold state and kept there. Food samples should not be frozen because many pathogens become much less active when they are.

The story is different if you are dealing with the biological material. They frequently deteriorate with time, so it is crucial to have a storage method (short and long term) that is effective and maintains sample integrity over time. Currently, billions of biological samples and specimens are stored in cold environments (refrigeration at −40 °C, low- and ultralow-temperature freezers at −50 to −800 °C) by researchers in academia, research institutions, hospitals, and commercial companies. The preservation and storage procedures for various samples are shown in Figures 5.2 and 5.3.

Inorganics, Metals, and Nutrients In Water and Wastewater

Test Parameters	Material Type	Preservation Conditions and Time
Acidity or Alkalinity	Plastic or Glass	Cool at 6 °C for 14 Days
Chlorine	Plastic or Glass	Analyze Immediately
BOD or Turbidity or Settleable Matter	Plastic or Glass	Cool at 6 °C for 48 Hours
Mercury	Plastic or Glass Rinsed with 1:1 HNO_3	1:1 HNO_3 to pH < 2 for 28 days
Chromium, Hexavalent	Plastic or Glass Rinsed with 1:1 HNO_3	Cool at 4 °C for 24 Hours
COD or Sulfate	Plastic or Glass	Cool at 6 °C for 28 Days
Chloride	Plastic or Glass	None for 28 Days
Lead	Plastic or Glass	HNO_3 to pH < 2 for 6 months
Surfactants	Plastic or Glass	Cool at 6 °C for 48 Hours
Nitrogen, Nitrate, Nitrite	Plastic or Glass	H_2SO_4 to pH < 2, Cool at 6 °C, 28 days

Figure 5.2: Preservation and storage procedures for various samples in water and wastewater.

5.4.1 Challenges in Using Cold Storage

There are some general difficulties that must be considered when chilling samples from various sources for examination:

- During cold packaging and transportation, a significant amount of waste is generated.
- As the number of samples collected increases, so do the costs of purchasing, maintaining, and operating cool refrigerators and freezers.
- Because of the heat produced by refrigerators and freezers, facilities must maintain ambient conditions at lower temperatures than would otherwise be necessary.
- Cold freezing can cause damage, particularly to biological tissues.
- Samples may be at risk of deterioration and loss in the event of a power outage or freezer malfunction.

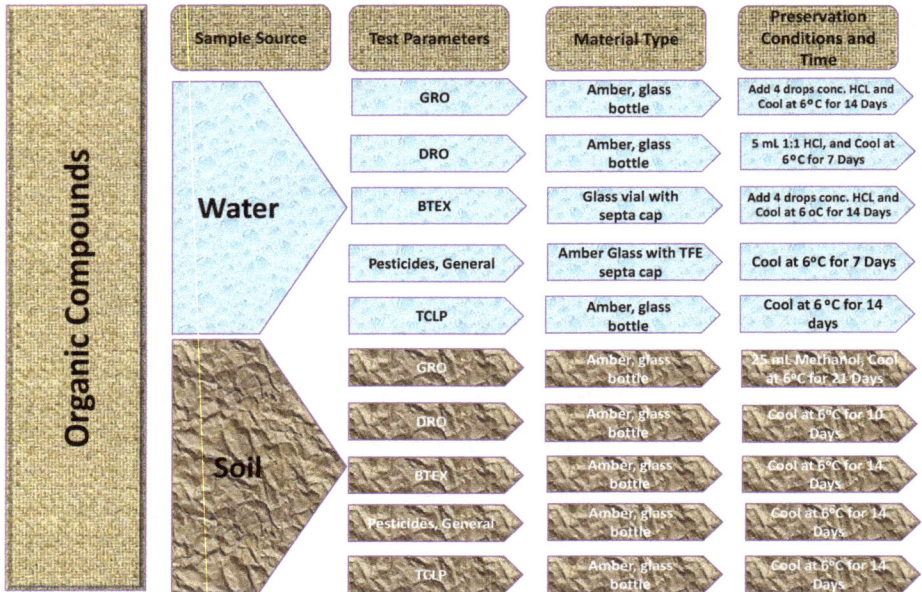

	Sample Source	Test Parameters	Material Type	Preservation Conditions and Time
	Water	GRO	Amber, glass bottle	Add 4 drops conc. HCL and Cool at 6°C for 14 Days
		DRO	Amber, glass bottle	5 mL 1:1 HCl, and Cool at 6°C for 7 Days
		BTEX	Glass vial with septa cap	Add 4 drops conc. HCL and Cool at 6 oC for 14 Days
		Pesticides, General	Amber Glass with TFE septa cap	Cool at 6°C for 7 Days
		TCLP	Amber, glass bottle	Cool at 6 °C for 14 days
	Soil	GRO	Amber, glass bottle	25 mL Methanol, Cool at 6°C for 21 Days
		DRO	Amber, glass bottle	Cool at 6°C for 10 Days
		BTEX	Amber, glass bottle	Cool at 6°C for 14 Days
		Pesticides, General	Amber, glass bottle	Cool at 6°C for 14 Days
		TCLP	Amber, glass bottle	Cool at 6°C for 14 Days

(left vertical label: **Organic Compounds**)

Figure 5.3: Preservation and storage procedures for various samples in water and soil.

5.5 Sample Types

There are four types of samples, which can be classified in the following sections.

5.5.1 Grab Sample

A grab sample (Figure 5.4) is a single sample taken at a specific time and location or a measurement taken at a specific time or over the shortest time period. Grab sampling is frequently carried out in an open bottle or container and can be done manually or by a suspension. A sample by hand usually requires the sampler to wade into the water to a predetermined place, take the sample while standing downstream, and then turn the open container upstream.

Grab sample is the most common sample type and the technique used by the vast majority of laboratories. A grab sample is taken when raw water or any other liquid or solution is collected in a beaker and tested for pH. It captures a snapshot of the characteristics of the sample at a specific point in time, so it may not be completely representative of the entire flow. Grab samples are ideal for small plants with low flow rates and limited staff that are unable to perform continuous sampling. In contrast, grab samples provide an immediate sample and are thus preferred for some tests.

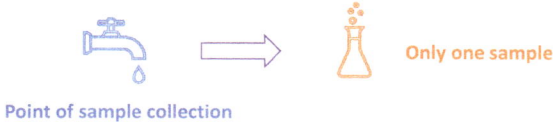

Point of sample collection

Only one sample is taken at the same time at the same location and handled, preserved, transported, and analyzed by one investigator

Figure 5.4: Grab sample.

Grab samples are preferable for some tests such as pH, dissolved oxygen, and total residual chlorine in water that can all change dramatically very fast once the sample is taken out of the flow.

Grab samples must be meticulously taken in order to be as accurate as possible in representing the water. When the plant is getting close to its typical daily flow rate during the day, it should be taken. The effluent sample should be taken after the raw water has fully completed the treatment process if grab samples are being used to measure plant efficiency by collecting samples of both treated and untreated water.

Whenever a grab sample is used in mining, the sampling process often entails selecting pieces of ore at random from muck piles, chutes, or the tops of ore cars. When carried out in this way, it is a haphazard procedure that should only be used to roughly estimate the grade of the ore and is frequently used just for that purpose.

When talking about the food industry, a grab sample – often taken with a sterile needle and of low volume – is also not gathered over time. It offers a "snapshot" of the process or outcome. For instance, a grab sample obtained from the pasteurizer just represents the product that is flowing through at that specific moment and not the entire run.

Grab sampling entails gathering a sample fluid in a pipeline, tank, or another system. It is also referred to as spot sampling, laboratory sampling, or field sampling. The sample is then tested to assist operators in verifying the operation of the process, assessing products for environmental pollutants in accordance with local laws, and determining whether the product meets client requirements. Grab sampling is a procedure that must be carried out using dependable tools and in compliance with accepted best practices. If the procedure is done wrong, the sample's integrity may be damaged, giving operators a faulty analysis of their systems that can have a negative impact on their decision-making.

Grab sampling is another method for taking a sample from a fluid system for distant laboratory analysis. Whenever you need to collect fluid for examination, including next to storage containers, along long transport lines, on process lines, at flare locations, and in other places, sampling panels can be deployed. And even though real-time process fluid analysis is a growingly popular feature of online analyzers, traditional grab sampling still provides a number of advantages.

Grab sampling systems are typical:

- More affordable than online analyzers
- Simple to install closer to process lines
- Easier to install and maintain than analyzers, which must be kept in analyzer shelters
- Enables all sample analyses to be carried out in a single laboratory
- Can be used to validate the results of an online analyzer

5.5.2 Composite Sample

Grab samples that were taken multiple times at the exact location are referred to as "composite samples." Continuous sampling or the blending of discrete samples is a method for collecting composite samples throughout time. A composite sample, also referred to as an integrated sample, is made up of a variety of separate grab samples that were obtained in proportion to the flow at each time point and gathered at regular and predetermined intervals. While composing, samples may be taken from just one area of units or containers, and the samples may be combined or blended to create a composite, which may then be sampled once and tested.

There are numerous available composite sample examples. Collecting pond water samples for 2 days into a single large container is one example. Another example is the composite samples of household dust, which provide useful information while necessitating the least amount of sample collection labor and financial investment in analytical tests. Composite samples, like grab samples, have advantages and disadvantages and are not suggested for all testing. These are summarized in Table 5.1. The ability of composite samples to consider variations in flow and other variables over time is their greatest strength. This aids the operator in obtaining a comprehensive understanding of all the impacts that the influent will have during the treatment process. On the hand tests of water characteristics that vary during storage, such as dissolved gases, or of water characteristics that change when samples are combined together, like pH, composite samples cannot be utilized.

The automatic sampling equipment used to obtain composite samples can be programmed to take a sample every so many hours for 1 or 2 days. The frequency can change based on the number of tests required, the size of the treatment facility, and the requirements for the permit. Figure 5.5, for example, depicts a composite sample taken every 8 h over the course of 2 days.

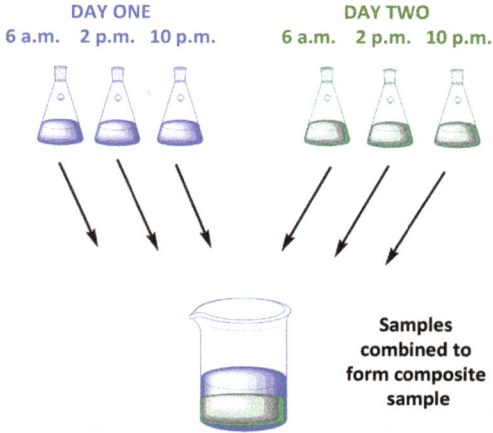

Several grap samples taken at different times
from the same location

DAY ONE
6 a.m.   2 p.m.   10 p.m.

DAY TWO
6 a.m.   2 p.m.   10 p.m.

Samples
combined to
form composite
sample

Figure 5.5: Composite sample.

Table 5.1: Composite sampling benefits and drawbacks.

Benefits	Drawbacks
Can lower the price of the analysis	Information on the individual samples that make up the composite is lost
More accurately calculates the mean concentration	Maximum concentrations result in information loss
Find the locations with the greatest contamination levels	Matrix homogenization challenges, for example, with clay soils

5.5.3 Duplicate Sample

To ensure that the sampling process is precise, a duplicate sample (Figure 5.6) is collected. It is investigated twice by two separate groups of researchers. Duplicate samples are typically collected in the form of two sample containers taken at the same time at the same location and handled, preserved, transported, and analyzed in the same way by the sampling and analytical procedures. They are used to compute the overall variance of the method, which includes sampling and analysis.

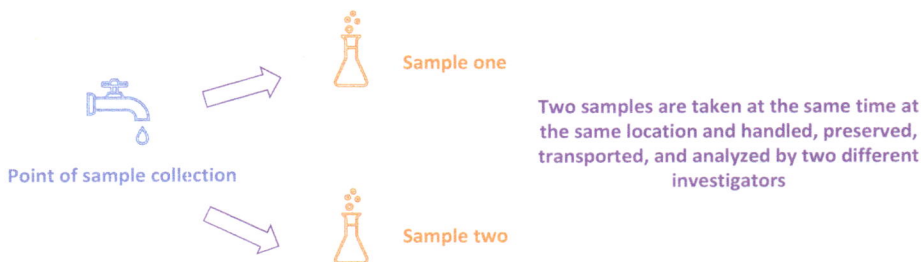

Figure 5.6: Duplicate sample.

5.5.4 Split Sample

A split sample used to assess the analytical performance is depicted in Figure 5.7. At least two batches of a single grab sample are typically separated, and each of which represents a representative portion of the original sample. This type of sampling is typically used when comparing test results between field kits and laboratories or between two laboratories. For comparison of sample batches in the laboratory, two aliquots of a single sample are created. One aliquot is examined using the primary test technique, and the second aliquot is examined at a different acceptable laboratory using the same or a different test method.

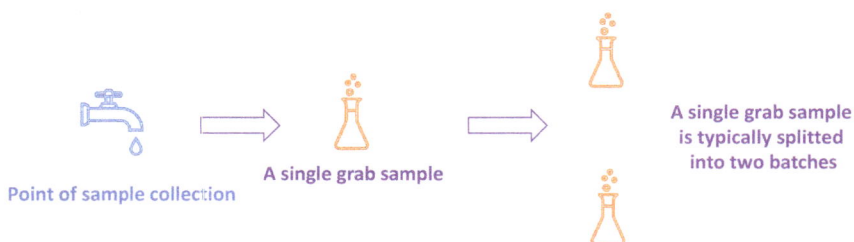

Figure 5.7: Split sample.

For any of the following objectives, any laboratory will advise and carry out a split sample:
- Verify the accuracy of unregulated analytes
- Compare a test or analyte to a different comparison approach
- Locate the cause of flawed testing procedures or test findings
- Verify a novel analytical technique

5.6 General Sampling Rules

The following general rules for sampling must be understood and followed by any researcher, lab technician, or chemist:

– For each given point of sampling site, samples must be taken from the least to the most contaminated sampling locations.
– While sampling, disposable latex gloves must be worn, and they must be brand-new and unused for every single sampling location.
– Use a basin and spatula to properly mix the samples when composing or combining solid materials.

5.7 Sample Handling

Between the sampling site and the laboratory, the sample is frequently handled with insufficient care. The sample integrity needs to be appropriately secured and conserved. During analytical testing, sample transfers must be recorded, including:

– the time;
– the goal and duties of the sample analysis;
– the code is used to identify a lab sample; and
– identification of the person and the location of storage where the sample is moved both to and from.

5.8 Chain of Custody

It is crucial to record who handled the sample, what duties each handler had at different points between the sampling site and the handler, as well as what steps the handler had taken to ensure sample integrity while the samples were in his/her care. In other words, the chain of custody needs to be upheld and verified. Figure 5.8 shows the minimum information that has to be in a chain of custody document.

The activities conducted on each specific sample, as well as any batch of samples taken for analytical testing, are documented in the *Laboratory Chain of Custody* paperwork. It is essentially a procedure that keeps track of how a sample is moved throughout the life cycle of collection, storage, and analysis by noting who handled the sample when it was collected or transferred, and for what reason.

There are two aspects to the chain of custody for any particular individual sample. The sample's custody from collection to delivery to the lab is tracked via the external chain of custody that was started at the collecting site. The internal chain of custody for the laboratory is from sample acceptance to sample disposal.

The internal chain of custody for the laboratory is documented, and this documentation may take the form of worksheets, logbooks, forms, or electronic records.

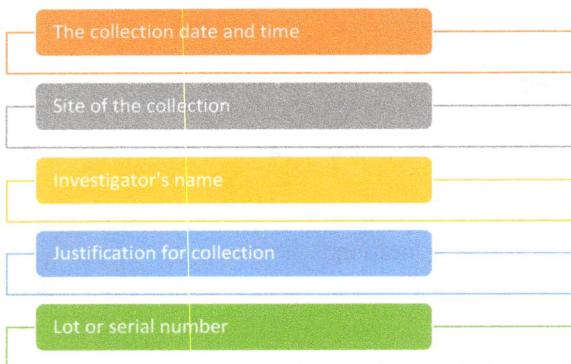

Figure 5.8: Chain of custody document information.

There must be an ongoing record of who handled the samples and where they were stored inside the laboratory's internal chain of custody. The date, place, action taken, and the person who performed the activity can all be included in the written or electronic record.

The authorized staff members involved in each entry into the internal chain of custody for the laboratory must do so. Samples are thought of as being in custody if they are:

- in the actual custody of approved laboratory personnel or another authorized individual;
- in the vicinity of a permitted laboratory and within the control zone; and
- kept in a place with strong security restrictions and designated storage.

5.9 Laboratory Sampling Procedures

In general, samples should be kept in a clean, dry, dark, cold, and ventilated environment. Food samples must be kept apart from other samples. Perishable items must be stored in refrigerators or freezers, and the storage temperature must be monitored on a regular basis.

Hands should be washed both before and after collecting medical samples, and appropriate personal protective equipment should be worn in accordance with established safety protocols when handling medical-related samples. Samples should also be taken in sterile containers with tight-fitting lids to avoid contamination and spillage.

5.10 Chemical Plant Sampling Procedures

The sampling procedure is determined by the type of chemicals handled by the chemical technician. The following are general guidelines for the chemicals:
- Before collecting samples, freeze the chemical cold pack.
- If the sample is taken from within the chemical plant, take it to a sampling point after treatment (if applicable) but before entering the distribution system.
- Remove any attachments, such as hoses, filters, screens, or aerators, from the sampling point.
- Fill out the laboratory form and sample label before collecting the sample. The laboratory forms vary, but the following minimum information must be included:
  - the sample type and purpose;
  - the date and time the sample was collected; and
  - sample location.
- Fill the sample container to the bottle's shoulder. If you are using a collapsible sample bottle, make sure to expand it before filling it by gently blowing it into the container's mouth.
- Keep the samples refrigerated until they are shipped.

5.11 Classical Analysis Methods

Chemical analysis is carried out by a group of professionals or researchers who have in-depth knowledge of special techniques like infrared (IR) absorption, emission spectroscopy, liquid chromatography, gas chromatography, or any other related procedures. Using both conventional wet chemical techniques and cutting-edge instrumental techniques, the sample analysis is made easy to handle, understand, and analyze. Prior to the analysis, traditional qualitative procedures such as precipitation, extraction, distillation, melting point, boiling point, and solubility are further techniques that can be used.

5.11.1 Qualitative Analysis

When doing a typical qualitative analysis, the analyte is combined with one or more chemical reagents. Analyzing chemical processes and the components that make them up can assist in determining the analyte. The reagents are selected in a way that ensures they will specifically react with one or a single class of chemical compounds to yield a unique reaction product. The end result of the reaction can occasionally be a precipitate, a vapor, or a colored substance.

Take copper(II) as an illustration, which interacts with ammonia to produce a rich blue copper–ammonia complex. Similar to this, yellow lead chromate is created when dissolved lead(II) combines with chromate-containing fluids. Positive ions and negative ions can both be qualitatively tested using the same technique. A typical instance is the creation of carbon dioxide gas bubbles as a result of the reaction between carbonates and strong acids.

Chemical interactions between newly added reagents and the functional groups of organic molecules are involved while conducting organic qualitative testing. This kind of reaction can reveal a portion of the organic molecule, but frequently not enough to fully classify the molecule.

Functional group analysis is used with other measurements, like boiling temperatures, melting points, and densities, to categorize the complete molecule. An excellent illustration of a chemical reaction that may be used to categorize organic functional groups is the interaction between bromine in a carbon tetrachloride solution and organic molecules having carbon–carbon double bonds. The unique reddish-brown hue of bromine is lost due to the bromine addition through the double bonds, revealing the presence of a carbon–carbon double bond.

5.11.2 Quantitative Analysis

Traditional quantitative analysis includes gravimetric and volumetric methods. Both methods rely on protracted chemical processes that involve the analyte and other reagents.

5.11.2.1 Gravimetric Analysis
When too much reagent is introduced and reacts with the analyte, a precipitate is created. After being cleaned, dried, and weighed, the filtrate is measured. Using its mass, the tested material's concentration or quantity in the analyte is determined.

5.11.2.2 Volumetric Analysis
Volumetric analysis is often referred to as titrimetric analysis. The reagent, the titrant, is added to the analyte gradually or in steps using a burette. The equivalence point of the titration, the point at which the quantities of the two reacting species are equivalent, which is typically evident as a color shift, is the key to performing a good titrimetric analysis.

5.12 Instrumental Analysis Methods

In reality, the descriptions of the structures of millions of organic compounds in the scientific literature are based on reliable experimental data. The following sections will provide examples of spectroscopic investigations and other characterization methods that significantly contribute to the strongest structural evidence. The methods outlined below produce information on specific distinguishing characteristics.

5.12.1 Nuclear Magnetic Resonance (NMR)

Nuclear magnetic resonance (NMR) spectroscopy), shown in Figure 5.9, is one of the most widely used techniques by chemists and biologists for identifying molecular structures. It is based on the NMR phenomenon, according to which the intramolecular magnetic field around an atom in a specific molecular structure changes the resonance frequency. As a result, different atoms within the same molecule may produce different resonance signals that the equipment can detect.

Samples for NMR studies should first be dissolved in a deuterated solvent and prepared in a thin-walled glass tube (NMR tube). The NMR tubes should then be fitted with spinners and placed in the NMR spectrometer. Different absorption spectral regions correspond to different physical processes occurring within the analyte. Absorption of energy in the radio frequency (RF) region in the presence of a magnetic field is sufficient to cause a spinning nucleus in certain atoms to shift to a different spin state. As a result, NMR spectrometry can be used to determine the number and type of nuclei present in the groups attached to the atom containing the nucleus of interest.

The most commonly used solvent for NMR measurements is chloroform-d ($CDCl_3$). Other deuterium-labeled compounds that can be used as NMR solvents include deuterium oxide (D_2O), benzene-d_6 (C_6D_6), acetone-d_6 (CD_3COCD_3), and DMSO-d_6 (CD_3SOCD_3). Because some of these solvents have electron functions and can act as hydrogen bonding partners, the chemical shifts of different groups of protons can differ depending on the solvent.

5.12.1.1 Proton NMR Spectroscopy (^{1}H-NMR)

To generate a proton NMR spectrum, a sample solution is placed in a uniform 5 mm glass tube and spun to average any fluctuations in the magnetic field and tube flaws. An antenna coil sends RF radiation into the sample at the appropriate energy level. A receiver coil surrounds the sample tube, and the output of absorbed RF energy is monitored by a computer and a specialized electrical equipment. To generate the NMR spectrum, the RF signal from the sample is observed as the magnetic field is changed or swept over a short area. An equally effective strategy is to change the frequency of the RF radiation while keeping the external field constant.

5.12.1.2 Carbon NMR Spectroscopy (^{13}C-NMR)

Although the effectiveness and efficacy of ^{1}H-NMR spectroscopy as a tool for structural investigation have been demonstrated, no information is obtained when significant amounts of a molecule lack C–H bonds, particularly in polychlorinated compounds like

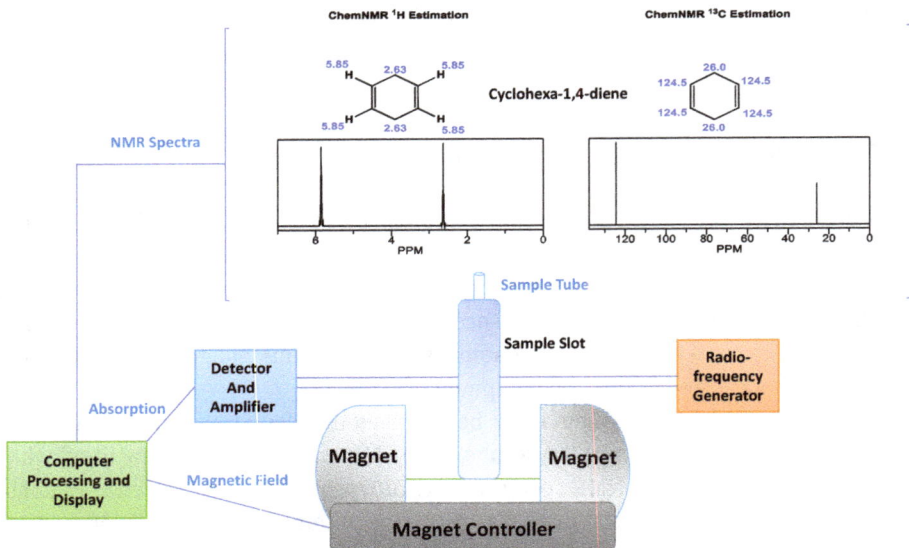

Figure 5.9: Nuclear magnetic resonance (NMR) spectrometer diagram.

chlordane. In these cases, ^{13}C-NMR is used to further elucidate the structure. Because there is no signal splitting and ^{13}C chemical shifts have a dispersion about 20 times greater than that of protons, it is more likely that each chemical shift is structurally distinct, and each carbon atom will emit distinct signal.

5.12.2 Atomic Absorption Spectroscopy (AAS)

Atomic absorption spectroscopy (AAS), is a spectroanalytical technique that exploits the optical radiation (light) absorption of free atoms in the gaseous state to quantify chemical elements. Their basic components are shown in Figure 5.10. The foundation of AAS is the light absorption of free metallic ions. The method is used in chemistry to establish the concentration of a particular element (the analyte) in a sample that will be subjected to analysis.

The AAS method consists of two steps:
- The sample atomization
- The absorption of the light from a light source by free atoms

The atomic absorption spectrometer has a light source that matches the narrow bands of light that a specific atom absorbs (a hollow cathode lamp), a flame or graphite furnace to heat the sample, a monochromator to select the wavelength of light, and a photodetector. An anode electrode and a hollow cathode made of the element being

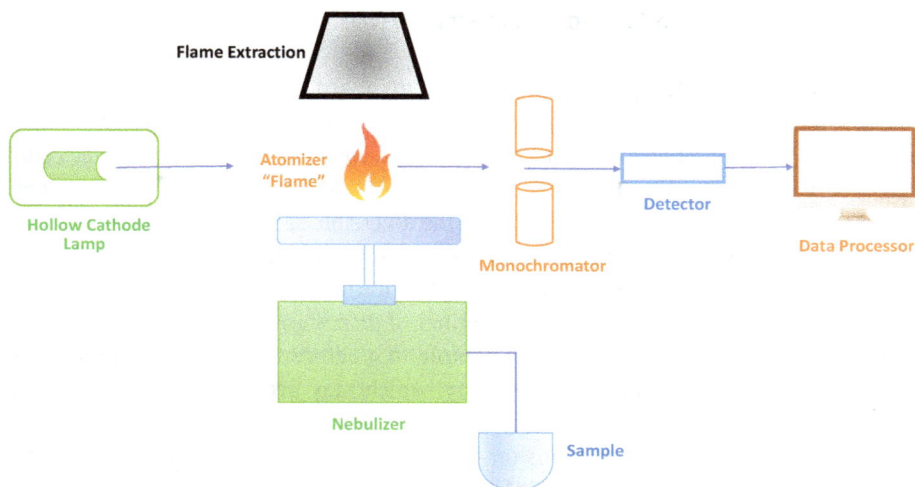

Figure 5.10: Atomic absorption spectroscopy (AAS) diagram.

analyzed comprise the hollow cathode lamp. Both were contained within a hollow tube filled with a noble gas. Gaseous ions bombard the cathode, causing metal ions to erupt.

Treatment with concentrated acids, such as HNO_3, HCl, or H_2SO_4, is a common sample preparation procedure for solid and viscous liquid samples. Following dilution of the treated solutions, samples can be injected directly into flame and graphite furnace AAS.

AAS, as a specialized method, offers a wide range of uses. Among those are biological and forensic analysis, environmental and marine analysis, geological analysis, and pharmaceutical analysis. Examples of biological samples include food samples and human tissue samples. AAS can be used to estimate the concentrations of different metals and other electrolytes in samples of human tissue. In the food sector, AAS examines crops, animal products, and animal feeds. The environmental and marine analysis includes different types of water analysis. Drinking water, wastewater, and seawater are just a few of the diverse issues covered by water analysis. In contrast to biological samples, the preparation of water samples is more influenced by rules than by the sample itself. There is a wide range of analytes that can be tested, and some of which include commonly lead, copper, nickel, and mercury. Mineral reserves and environmental studies are both incorporated into the geological analysis. One example is testing lake and river sediment for lead and cadmium. The method can also be used to test for trace metal impurities and to assay a number of frequently occurring elements in medicines.

5.12.3 Ultraviolet-Visible Spectrophotometry (UV-Vis)

UV-Vis (ultraviolet-visible) spectroscopy is the study of absorption and reflectance spectroscopy in the UV and visible parts of the electromagnetic spectrum. This implies that it makes use of visible and nearby light. This light's relatively high energy causes electronic activation upon absorption. Wavelengths between 200 and 800 nm absorb only when conjugated pi-electron systems are present in the accessible section of this area.

It is a widely used technique in many fields of science, ranging from drug identification and nucleic acid purity checks and quantitation to beverage quality control and chemical research. It compares the number of discrete wavelengths of UV or visible light absorbed or transmitted by a sample to a reference or blank sample. The sample composition influences this property, which may provide information about what is in the sample and at what concentration.

While there are many variations on the UV-Vis spectrophotometer, consider the main components shown in Figure 5.11 to gain a better understanding of how a UV-Vis spectrophotometer works.

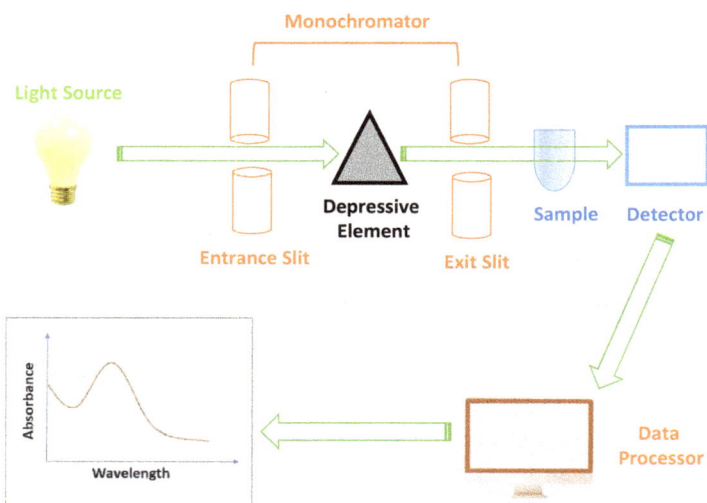

Figure 5.11: Main components of UV-Vis spectrophotometer.

5.12.3.1 Pros and Cons of UV-Vis Spectroscopy
UV-Vis spectroscopy is not an exception to the rule that no one approach is ideal. Nonetheless, the method has a few key benefits that are outlined below and contribute to its popularity:

- The method is nondestructive, enabling the reuse of the sample or moving on to further processing or analysis.
- Measurements are made rapidly, making it simple to integrate them into experimental techniques.
- Instruments do not require much user training before use and are simple to operate.
- Data analysis typically only necessitates minimal processing, which again means that user training is not overly complicated.
- The device can be used by many laboratories because it is often affordable to buy and maintain.

Despite the overwhelming advantages of this method, there are some drawbacks as well:
- Although wavelength selectors in actual instruments are not infallible, some light from a variety of wavelength ranges may still be transferred from the light source.
- In liquid samples, suspended materials frequently produce light scattering, which can result in significant measurement errors. Bubbles in the cuvette or sample will scatter light and produce inconsistent findings.
- Interference is from various absorbent parts.
- Any of the instrument's parts that are positioned incorrectly, notably the cuvette that is holding the sample, could produce results that are unreliable and erroneous. As a result, it is crucial that each component of the instrument is positioned consistently for each measurement and is oriented in the same direction.

5.12.3.2 Applications of UV-Vis Spectroscopy
Many applications and circumstances include, but not limited to,
- analysis of nucleic acids and modified nucleic acids;
- analysis of biopharmaceuticals;
- characterization of nanoparticles of extreme smallness;
- analyzing certain protein structural alterations;
- analyzing the authenticity of the food; and
- air quality surveillance.

5.12.4 Infrared (IR) Spectroscopy

IR spectroscopy studies the electromagnetic spectrum's IR region, which includes light with a longer wavelength and lower frequency than visible light. The study of a molecule's interaction with IR light is known as IR spectroscopy. IR spectroscopy can be studied in three ways: reflection, emission, and absorption. The primary application of IR spectroscopy is to determine the functional groups of molecules, which is important in both organic and inorganic chemistry.

The IR light frequencies that a molecule absorbs are measured using IR spectroscopy. These particular light frequencies are absorbed by molecules because they coincide with the frequency at which the bonds in the molecule vibrate. The typical IR absorption range for covalent bonds, shown in Figure 5.12, is 600–4,000 cm^{-1}, where the spectral regions that various types of bonds typically absorb are shown.

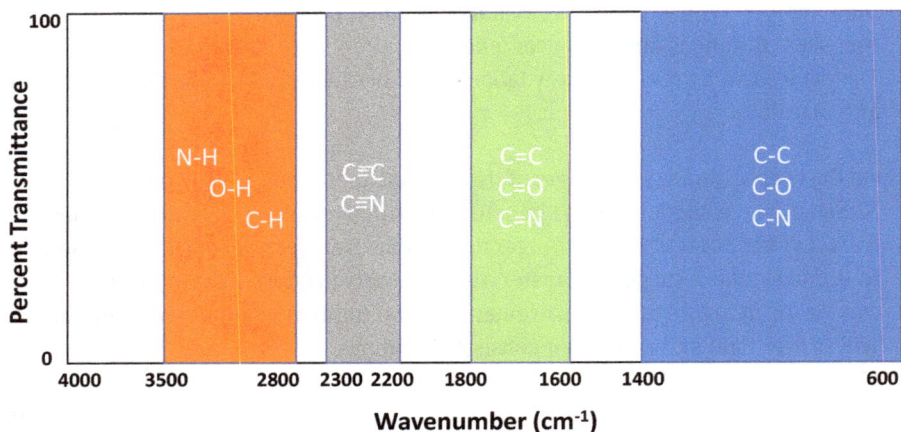

Figure 5.12: The typical IR absorption range.

An illustration of the IR spectroscopy apparatus is shown in Figure 5.13. The IR light source, a sample holder, a mechanism to divide the light into its component wavelengths, and a detector make up the fundamental parts of a spectrophotometer. The mechanism directs the light source's electromagnetic energy toward the material sample.

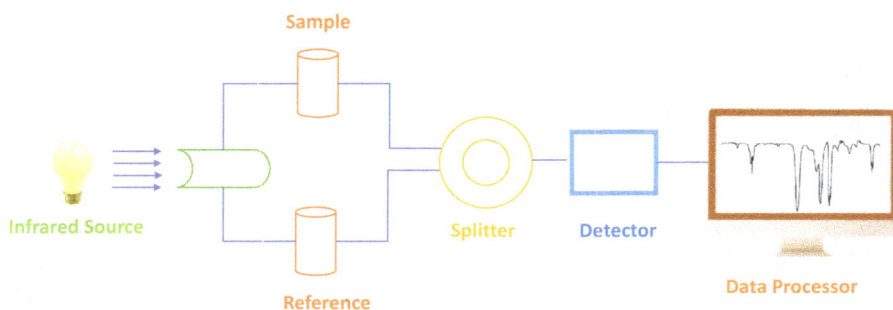

Figure 5.13: Main components of IR spectrophotometer.

IR spectroscopy can be used to analyze a variety of substances. They can exist in one of three states: solid, liquid, or gas. By crushing the sample with a mulling agent that has an oily texture, solid samples can be created. This mull can now be spread thinly on a salt plate for measurement. On the other hand, liquid samples are typically held between two salt plates when being measured. Sodium chloride, calcium fluoride, or even potassium bromide can be used to make salt plates. The sample cell for gaseous samples, which are often measured in parts per million, must have a comparatively long path length, meaning that light must travel a relatively far distance inside the sample cell.

Although IR can tell whether a certain functional group is present or absent and can create a molecular fingerprint that can be used to compare samples, it cannot give precise information or evidence of a molecule's formula or structure. It offers details on molecular pieces, particularly functional groups. As a result, it has a relatively narrow field of application and needs to be combined with other methods to produce a more comprehensive image of the molecular structure.

5.12.5 Mass Spectrometry (MS)

By examining variations in charge-to-mass ratios (mass/charge; m/z), mass spectrometry (MS) is an analytical method that can be used to measure the molecular weight of ionized particles such as atoms and molecules. High-energy electrons are what ionize the sample molecules. Electrostatic acceleration and magnetic field perturbation give a precise molecular weight by measuring the mass-to-charge ratio of these ions. The molecular ion's structure might be related to the patterns of ion fragmentation.

A mass spectrometer's three primary purposes and the corresponding parts are:

- a little amount of a molecule is ionized by the loss of an electron, typically to cations (the ion source);
- according to their mass and charge, the ions are separated and sorted (the mass analyzer); and
- thereafter, the separated ions are identified and counted, and the results are shown on a chart (the detector).

Ion source, mass analyzer, and detector are shown schematically in Figure 5.14 as the three primary parts of a mass spectrometer. The material that will be examined is ionized by the ion source. The ions are then moved by magnetic or electrical forces to the mass analyzer and detector, where they are detected. The signals from these devices are then recorded and transformed into digital data to create mass spectra.

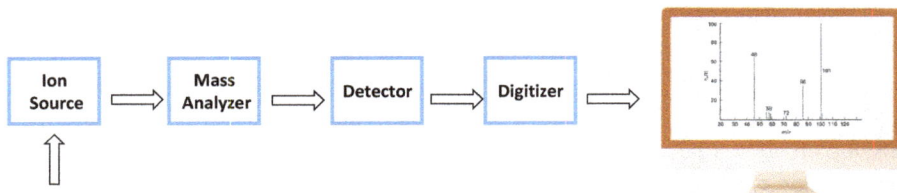

Figure 5.14: Main components of mass spectrometer.

5.12.6 X-Ray Fluorescence (XRF)

The elemental makeup of a substance can be ascertained using the nondestructive analytical method known as X-ray fluorescence (XRF). XRF analyzers may assess the chemistry of a sample by measuring the fluorescence X-ray, or secondary X-ray, that a sample emits when triggered by a main X-ray source. In many analytical labs across the world, including those that deal with metallurgy, forensics, polymers, electronics, archaeology, geology, and mining, XRF is used as a fast characterization technique.

According to the wavelength and intensity of incident X-rays, the XRF has the potential to detect X-ray emission from practically all elements; however, it is ineffective for lithium, beryllium, sodium, magnesium, aluminum, silicon, or phosphorus.

Gold, silver, and platinum group metals, as well as nonprecious alloying metals, impurities, and gold plating, can all be nondestructively analyzed using XRF. Even some false gemstones like cubic zirconia, titanite, and leaded glass can be recognized by XRF. It may also detect potentially dangerous heavy metals in automobiles, medical equipment, and electronics before they are put on the market.

As shown in Figure 5.15, modern systems primarily consist of three parts: an X-ray source, a detector, and a signal processing unit. XRF can be thought of as a straightforward, three-step atomic-level process. An electron in an atom of the substance is removed by primary X-rays from one of the orbitals that surround the nucleus.

A hole forms in the orbital, putting the atom in an unstable, high-energy state. To restore balance, an electron from an outer orbital with greater energy falls into the hole. This is a lower energy state, and the extra energy causes the production of fluorescent X-rays. As a result of the energy difference between the expelled and replacement electrons, which is distinctive of the atom in which the fluorescence process is taking place, the energy of the fluorescent X-ray that is released is directly tied to a specific element that is being studied.

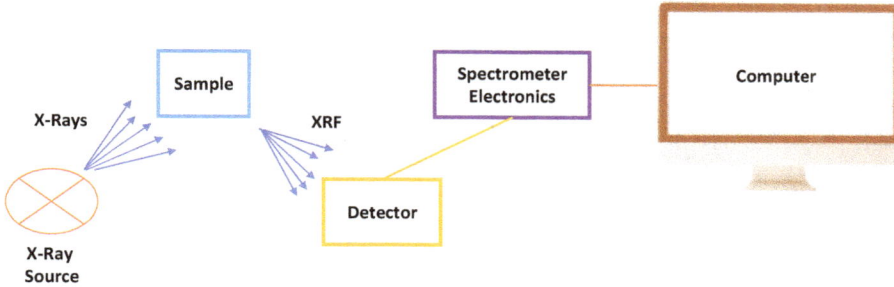

Figure 5.15: Main components of XRF.

5.12.7 X-Ray Diffraction (XRD)

X-ray diffraction (XRD) is a popular technique for identifying the crystallinity and structure of solid samples. The two distinct types are single crystal and powder XRD methods. The method is based on the phenomenon known as crystal XRD, which is caused by X-ray scattering by electrons of atoms present in the sample without changing their wavelength.

X-ray beams are utilized because their wavelength is equivalent to the distance between atoms in the sample, as opposed to employing much longer wavelengths, which would be unaffected by the distance between atoms. This implies that the atoms' spacing within the sample will have an impact on the angle of diffraction. The following can be accomplished by utilizing XRD techniques:

– calculate the typical distances between atom layers and rows of a substance;
– establish the direction of a specific grain or crystal;
– assess the internal stress, size, and form of tiny crystalline regions; and
– find the crystal structure of a substance you do not know.

In the fields of geology, engineering, biology, electronics, and pharmaceuticals, the use of XRD for the identification of unknown crystalline materials, such as inorganic compounds or minerals, is crucial.

Figure 5.16 depicts the three basic components of the XRD instrument: the X-ray detector, sample holder, and X-ray source.

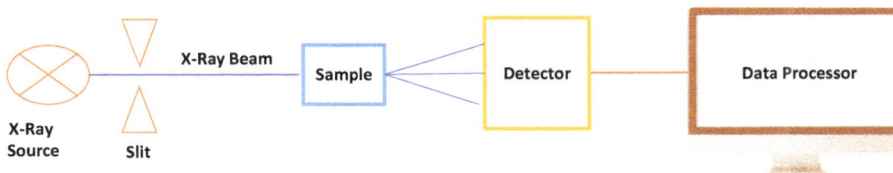

Figure 5.16: The three basic components of the XRD instrument.

XRD has many advantages, such as the ability to identify unknown minerals and materials, the requirement for rapid, precise sample preparation prior to analysis and the simplicity with which the resulting data can be interpreted, but it also has some disadvantages, such as sample homogeneity, access to standard reference data, and the fact that sample preparation frequently entails grinding the samples to a powder.

5.12.8 Energy-Dispersive X-Ray Spectroscopy (EDX or EDS)

Energy-dispersive X-ray spectroscopy (EDX or EDS) is a chemical technique utilized to analyze materials. It works in conjunction with an electron microscope, such as a transmission electron microscope (TEM) or a scanning electron microscope (SEM), to visualize the internal structure of solids or study the surfaces of solid objects. Figure 5.17 depicts the EDX system, which consists of X-ray source, a sensitive X-ray detector, and data processor for gathering and analyzing energy spectra.

There is no particular sample preparation needed for EDS other than what is needed to image the sample in the SEM or TEM for qualitative analysis; nevertheless, the sample needs to be bulky, flat, and polished for quantitative analysis in the SEM.

EDS can quickly gather qualitative chemical data, determine a surface composition in a semiquantitative manner, map the lateral distribution of chemical elements, and create compositional profiles over a surface. All stable elements, with the exception of lithium, hydrogen, and helium, can be found.

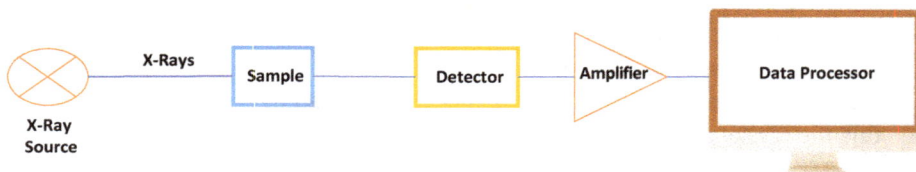

Figure 5.17: Main components of EDX or EDS instrument.

5.13 Questions

5.13.1   What are the six major activities that determine data quality?

5.13.2   List down the benefits of sampling in chemistry laboratories.

5.13.3   What kind of vessel or container material can be used to collect inorganic materials, trace metals, extractable organics, volatile organics, and microbiological samples?

5.13.4   What are the challenges in using cold storage?

5.13.5   List down the names of the four types of samples.

5.13.6   What are the composite sampling benefits and drawbacks?
5.13.7   What are the basis of the NMR spectroscopy?
5.13.8   What are the most commonly used solvents for NMR measurements?
5.13.9   What are the main UV-Vis spectroscopy applications?
5.13.10 List down the main XRF components.

Chapter 6
Chemical Laboratory and Chemical Plant Safety Procedures

6.1 Introduction

Regardless of the working environment, safety is always a top priority. Chemical operators and technicians must be aware of the potential safety risks present in their units and take all reasonable steps to keep both themselves and others safe. Doing the following steps puts safety first:

– Chemical plant operating procedures should outline what must be done to maintain operations' efficiency and safety. These protocols outline the proper way to use the equipment, including a list of safety measures to take and offer guidance on how to handle crises. For instance, a plant's operating guidelines may stipulate that whenever certain duties are carried out, breathing protection must be worn.

– Technicians are often in charge of carrying out processes that contribute to the security of workers in chemical plants. For instance, chemical plant operators or technicians must position the equipment or its system safely when doing maintenance on plant equipment. This can be achieved by separating the machinery and adhering to protocols that prevent accidental equipment starts while repairs are being made.

– Housekeeping is a crucial component of safety. By reducing or eliminating specific forms of fire hazards, good housekeeping can reduce or even eliminate the risk of slips, trips, and falls as well as making fire prevention easier.

– There is always a chance that anything could go wrong despite diligent attempts to prevent mishaps. Chemical plant operators and technicians must be prepared to respond to accidents and crises at all times.

– Chemical plant operators and technicians need to be familiar with their facility's emergency procedures in order to be properly prepared.

– Operators and technicians of chemical plants must also be aware of the dangers posed by the materials used in the facility and know where hazardous items are kept and used.

– Chemical plant personnel and technicians must understand how to use personal protective equipment to avoid exposure to dangerous compounds.

– Chemical plant operators and technicians must be familiar with the locations of the fire alarm boxes and the fire-fighting supplies in order to be prepared for a fire. Also, they must be able to react appropriately in case of plant fires.

https://doi.org/10.1515/9783111191492-006

6.2 Safe Usage of Hazardous Chemicals in the Laboratory and Plant

In the laboratory, technicians often handle a variety of substances. A chemical's level of danger is influenced by both its composition and its use. We must consider chemical's properties when deciding how hazardous it is. The majority of chemicals fall into the following groups:
- Chemicals that are flammable have the ability to ignite quickly and burn in the air. Solids, gases, and liquids all have the potential to ignite and explode. Examples include laboratory solvents, certain adhesives, and paint thinner.
- Corrosive chemicals are those that have the ability to directly corrode metal or burn, irritate, or damage living tissue. Strong bases and acids, as well as oxidants and dehydrating agents, fall under this group. Examples include hydrogen peroxide, potassium, ammonium, and sodium hydroxides as well as sulfuric, nitric, and hydrochloric acids.
- When mixed with heat, light, water, or air oxygen, oxidizers, and reactive compounds react violently, resulting in explosions or violent chemical reactions. Examples include nitrates, chlorates, nitrites, peroxides, crystallized picric acid, ethyl ether, and metals that react with water such as sodium.
- Compressed gases that are stored under high pressure might hurt people nearby physically if the tanks and valves that control them develop cracks or other damage. Examples include oxygen, propane, nitrogen, chlorine, and other commonly used compressed gases.

6.2.1 Laboratory Use of Flammable Chemicals

The number of flammable chemicals or other combustible materials near the handling area should be kept to a minimum during all laboratory procedures involving flammable chemicals in order to prevent the travel or accumulation of flammable vapors, minimize their release, and eliminate sources of ignition. While handling flammable chemicals, the following safety precautions should always be taken:
- Employ fume hoods wherever practical, especially when moving or heating volatile substances.
- Use combustible gases in a fume hood at all times.
- Unless it is part of an experimental technique that has been given, never use open flames in the same area where flammables are being used.
- In places where flammable vapors are anticipated to exceed 10% of the lower flammability limit, control other sources of ignition and heat in the lab such as electric motors and ovens.
- Use only electrical equipment, such as heating and stirring plates, that is designated as explosion-proof.

- Ground the metal container before transferring flammable substances from one to the other.
- Reduce the amount of dust produced when handling combustible solids.
- Ensure that the area around the operation has the appropriate extinguishing media such as class D powder for combustible metals.
- Never let the operations of the solvent distillation run unsupervised.

6.2.2 Laboratory Use of Corrosives

While working with corrosive compounds, always follow these safety precautions:
- Just what is required should be purchased; small quantities are advised for simpler handling and storage.
- For transporting chemicals between floors, bottle carriers or some other type of containment should be employed.
- Put on the appropriate safety equipment.
- Never add chemicals to water all at once; instead, use concentrated acid.
- Inorganic acid spills should be kept away from ignition sources because they can react with metals to create explosive hydrogen gas, and glacial acetic acid can catch fire since it is an organic acid and hence a combustible chemical.
- The use of perchloric acid necessitates a written, certified protocol. Never add a concentrated acid to a base or a concentrated base to an acid when neutralizing corrosives. The number of people utilizing the acid should be kept to a minimum, and all users should be aware of its chemistry, dangers, recommended handling practices, and emergency protocols. Perchloric acid must never be heated. Disposal protocols should be established before doing research with perchloric acid.
- Hydrofluoric acid is an immediate poison. There must be a written, approved protocol before handling it. To prevent preparing solutions, acid should ideally be obtained at the concentration that will be used. Always use with the sash as low as feasible and no higher than 15 inches in a working fume hood. Use rubber or neoprene gloves, a face shield that protects your face and neck, chemical splash goggles, nonabsorbent-resistant clothing, and a rubber or neoprene apron. After each use, carefully wash your hands. Employ only durable equipment such as Teflon and polyethylene. Disposal protocols should be established before conducting hydrofluoric acid tests.

6.2.3 Laboratory Use of Reactive Chemicals

Always keep in mind the following safety measures when working with reactive materials:

- Look into the reactive chemical's purity. Assess whether the experiment will be more dangerous if there are contaminants or results of spontaneous breakdown like peroxides.
- Carry out small-scale pilot tests to evaluate the reaction's physical and thermodynamic characteristics.
- Employ a diluted solution or as little of the reactive chemical as you can.
- Evaluate all available tools for regulating reaction variables. Both the adding rate and the rate at which the energy for activation is provided are controllable. Cool exothermic reactions effectively to regulate the rate of the reaction. If necessary, do not forget to include cooling options for both the liquid and vapor stages. When adding chemicals to pressurized systems, pressure relief valves should be present and tested.
- Choose the appropriate agitation level and mixing speed. Add oxidants gradually while cooling or mixing appropriately.
- When necessary, wear a face shield in addition to goggles.
- Work in a fume hood while shielded from harm by the sash.
- Have the necessary emergency supplies nearby.
- Alert lab personnel to any unusual dangers that could be brought about by the usage of a reactive chemical.

6.2.4 Laboratory Use of Compressed Gas Cylinders

The following safety precautions should always be kept in mind when handling compressed gas cylinders:
- Gas cylinder handling, use, and attachment of regulators must adhere to the manufacturer's instructions.
- Avoid using flammable gases close to exit routes.
- Use protective goggles or glasses when putting or taking out gas cylinder regulators.
- Connect the appropriate regulator made for the specific gas being utilized. Cylinder valves have been standardized for a few distinct gas families in order to prevent the exchange of regulator equipment between gases that are incompatible.
- After attaching the regulator, make sure all hose connections are tight with clamps, tie down any loose hoses to prevent them from moving suddenly when pressure is applied, and, if necessary, install a trap between the regulator and the reaction vessel to stop backflow.
- After installing or replacing the regulator, make sure it is firmly in place by looking for leaks.
- Every gas line that exits a compressed gas source needs to be clearly marked. Where flammable compressed gases are present, signs should be prominently displayed.

- Unless the cylinder is specifically made for such use and is so labeled, never combine gases in a cylinder.
- Empty cylinders should have the regulator taken off and the protector cap put on, and they should be marked "MT" or "EMPTY" with the lab manager being made aware of this.
- Acetylene pressure gauges ought to show a warning red line. Acetone has been used to dissolve the acetylene gas in cylinders, which must always be used upright. A cylinder should not be used until it has been upright for at least 30 min after being handled or stored in a nonupright posture. A flash arrester is required on an acetylene cylinder's outlet line. Copper, among other tube materials, can create explosive acetylides.
- Organic materials, like oil or grease, will quickly oxidize under pressure when exposed to oxidizers like oxygen and chlorine, leading to an explosion. On valves or gauges designed for cylinders holding oxidizers, never use grease or oil.

6.3 Personal Protective Equipment

It is important to remember that you are responsible for your own safety and that of others. Personal protective equipment consists of lab coats and gowns, shoe covers, and respirators. Figure 6.1 provides a summary of the typical ones.

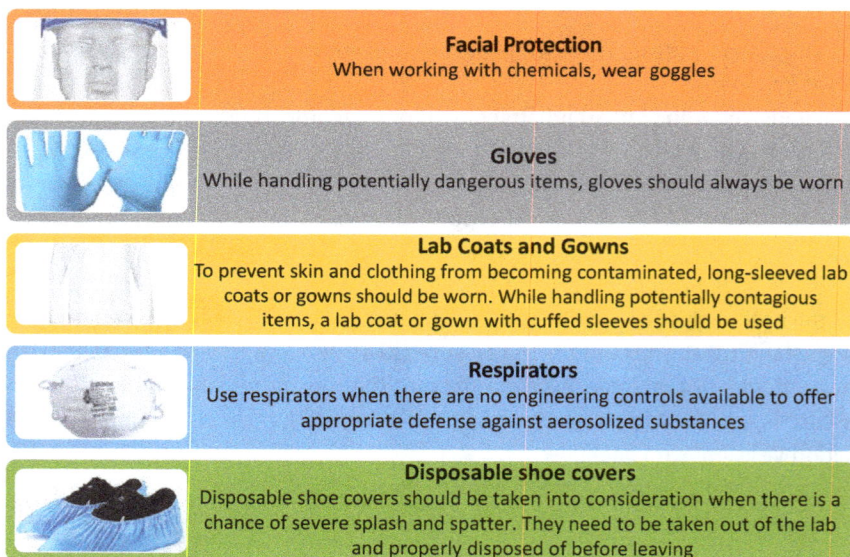

| **Facial Protection** |
| When working with chemicals, wear goggles |

| **Gloves** |
| While handling potentially dangerous items, gloves should always be worn |

| **Lab Coats and Gowns** |
| To prevent skin and clothing from becoming contaminated, long-sleeved lab coats or gowns should be worn. While handling potentially contagious items, a lab coat or gown with cuffed sleeves should be used |

| **Respirators** |
| Use respirators when there are no engineering controls available to offer appropriate defense against aerosolized substances |

| **Disposable shoe covers** |
| Disposable shoe covers should be taken into consideration when there is a chance of severe splash and spatter. They need to be taken out of the lab and properly disposed of before leaving |

Figure 6.1: Personal protective equipment.

6.4 Permits and Regulations

In a chemical laboratory or chemical plant, chemical technicians are often involved in different work environments, and some of them include working with dangerous chemicals or biohazards. Such cases need to be done under certain regulations and with special permits. Most chemical laboratories and chemical plants issue written work permit that authorize potentially dangerous work to be done during normal operating conditions. The information on a typical permit includes a description of the work to be done, special safety instructions regarding hazards, and signatures approving the work. The following are some of the cases where work permits are required from concerned supervisors or departments:

–   When working with lead compounds, mercury compounds, asbestos, benzene, vinyl chlorides, acrylonitrile, silica, coke oven emissions, ethylene oxide, isocyanates, and asbestos-related construction, a designated substance permit is necessary. Each substance requires special storage and handling.
–   Specific control measures must be in place before a permit can authorize the use of a designated substance overnight or unattended.
–   Specific control measures must be in place before the use of cryogenics.
–   Dealing with radioactive or biohazardous materials requires specific permission and particular training.

6.5 Global Regulations for Workplace Safety

Besides permit procedures, chemical technicians must also comply with other local or international standards and regulations. As a result of numerous government requirements over the years, chemical facilities have introduced safety precautions; many of these safety measures have an impact on chemical technicians. The following are three international laws that directly impact chemical technicians at various workplaces:

–   The hazard communications standard (HAZCOM or HCS).
–   The hazardous waste operations and emergency response standard (HAZWOPER).
–   Title III of the superfund amendments and reauthorization act (SARA III).

6.5.1 The Hazard Communications Standard

Hazard communications standard is another name for employee right-to-know (HAZCOM or HCS). It ensures that employees have access to information about any potentially dangerous substances used in the chemical plant. There are five main parts to the Hazard Communication Program. Among these include a written plan, employee education, Safety Data Sheets, chemical labeling, and chemical inventory.

All types of chemicals in the workplace, including liquids, solids, gases, vapors, fumes, and mists, are covered by the Hazard Communication Standard. Chemical producers and importers are required by HCS to classify their products and identify whether they constitute a physical hazard, such as being flammable or explosive, or a health hazard, such as being toxic or poisonous. If they are recognized as such, the manufacturer or importer must produce a Material Safety Data Sheet (MSDS), which is a document that provides more detailed technical information on hazardous chemicals and serves as a reference for workers, employers, emergency personnel, and health professionals. Chemical labels, which briefly describe the hazards associated with the chemical, are also required.

The three main objectives of HAZCOM are to:

– ensure that containers containing dangerous chemicals are appropriately labeled;
– notify staff members about dangerous chemicals and provide them with the necessary training; and
– guarantee the upkeep of a list of the chemicals utilized in a unit.

All chemical containers that enter or exit work areas are required to be labeled with particular information about the chemical within, according to the primary objective. The identification of the chemical, the name and address of the manufacturer, and any pertinent cautions regarding the chemical, such as its corrosiveness, must all be included on the label. Several danger identification techniques may be used in chemical factories and laboratories. Color and number combinations are frequently employed to indicate the contents of storage tanks and chemical containers. As shown in Figure 6.2, this hazard identification system uses colors to denote the type of hazard and numbers inside the colored areas to denote the severity of the hazard.

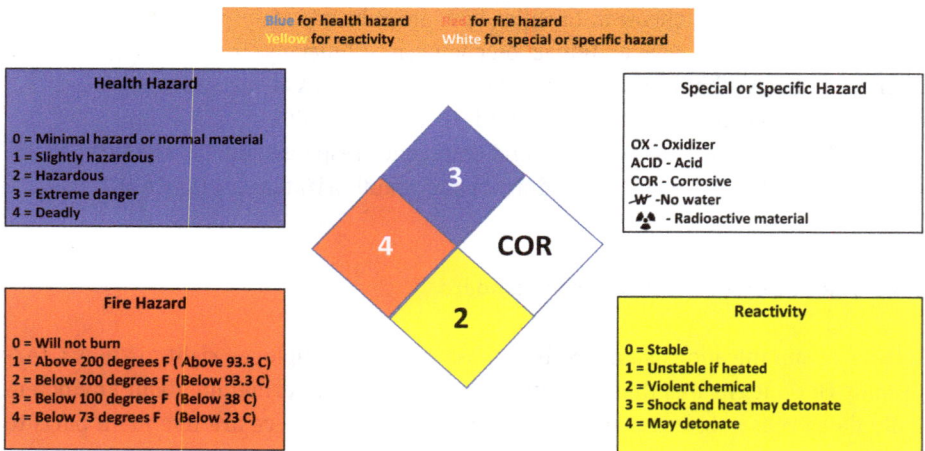

Figure 6.2: Different types of hazard identification systems.

The second objective of HAZCOM is to inform staff members about hazardous substances. The MSDSs indicate where to get resources for details on chemical dangers. The MSDS must contain the information indicated in Figure 6.3.

Figure 6.3: Material safety data sheet.

To adhere to HAZCOM's third objective, companies, academic institutions, and research facilities must keep a list of the chemicals used on their property. It is necessary to post and write the chemical inventory in a language that the departmental or laboratory staff can understand. Figure 6.4 displays the different benefits that a precise chemical inventory provides.

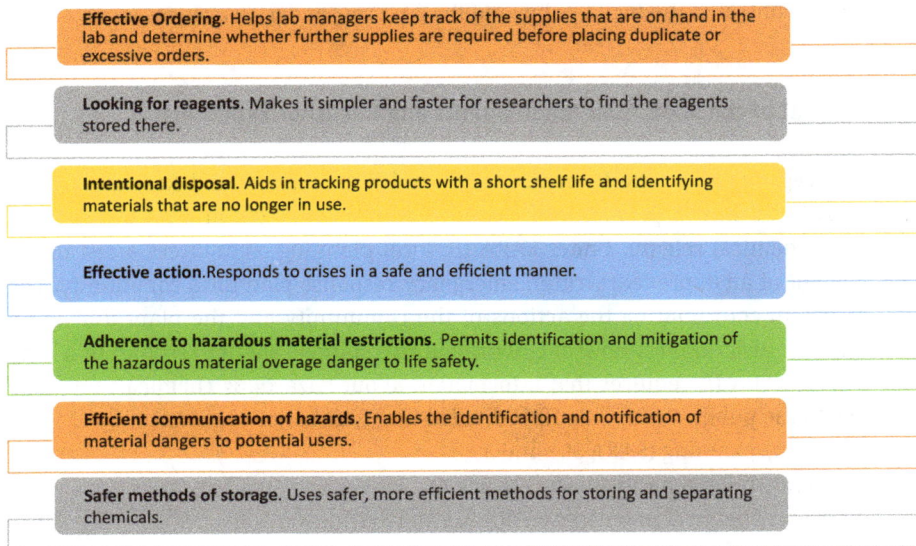

Figure 6.4: Different benefits that a precise chemical inventory provides.

6.5.2 The Hazardous Waste Operations and Emergency Response Standard (HAZWOPER)

Another regulation that affects chemical companies is the HAZWOPER standard. HAZWOPER is concerned about the potential for chemical spills. When harmful chemicals are handled, stored, or disposed of, it helps to ensure worker safety. Hazardous worker regulations are developed and updated by the Occupational Safety and Health Administration (OSHA). HAZWOPER includes the management of hazardous waste as well as emergency response, treatment, storage, and disposal.

HAZWOPER categorizes employees who may contact chemicals into five emergency response levels and call for different training at each level.

– Level 1 is the first responder awareness level. A Level 1 person is trained to detect a hazardous substance release and initiate an emergency response sequence.
– Level 2 is the first responder operations Level. A Level 2 person is trained to respond defensively to a potential or actual hazardous substance release.
– Level 3 is the hazardous materials technician level. A Level 3 person is trained to respond to releases or potential releases to stop the release.
– Level 4 is the hazardous materials specialist level. A Level 4 person with specific knowledge of hazardous substances is trained to support hazardous materials technicians.
– Level 5 is the on-scene incident commander level. A Level 5 person is trained to assume control of the incident beyond the first responder awareness level.

6.5.3 Title III of the Superfund Amendments and Reauthorization Act

The third regulation is Title III of the Superfund Amendments and Reauthorization Act (SARA III). SARA III promotes effective community reaction and planning in the event of a chemical release. Under SARA III, local planning committees assist in the development of an approved, written emergency response plan for a community. The plan assists in coordinating the actions of the community and the plant during an emergency. It also outlines how neighborhood fire stations should respond to a dangerous spill. SARA III requires that information about hazardous chemicals be made available to the public. SARA Title III is divided into four major sections:

– Emergency planning (SARA 302 and 303)
– Emergency release notification (SARA 304)
– Hazardous chemical inventory (SARA 311 and 312)
– Toxic chemical release inventory (SARA 313)

6.6 Questions

6.6.1 What are the safety precautions that should always be taken while handling flammable chemicals?

6.6.2 Give two examples of flammable chemicals.

6.6.3 Define corrosive chemicals.

6.6.4 Give four examples of compressed gases.

6.6.5 What are the personal protective equipment that are usually needed in the chemical laboratory?

6.6.6 What are the three main objectives of HAZCOM?

6.6.7 Briefly explain the hazard identification system.

6.6.8 What are the benefits that a precise chemical inventory provides?

6.6.9 How many emergency response levels does HAZWOPER have?

6.6.10 What are the four major sections of SARA Title III?

Chapter 7
Good Laboratory Practice (GLP)

7.1 Introduction

There are some fundamentals of laboratory work such as sample handling and laboratory behavior. Laboratories are not known for being the healthiest work environments. The presence of volatile hazardous organic liquids, explosive chemicals, and laboratory equipment does not guarantee optimal working conditions. However, it is possible to organize the laboratory environment in such a way that normal work can be performed without causing health problems.

Laboratory analyzers and attendants should go by a few standard guidelines to keep the area secure, healthy, and trouble-free. The term "Good Laboratory Practices" (GLPs) is used to describe them. Through nonclinical safety tests ranging from physiochemical properties to acute to chronic toxicity tests, GLP is a quality management system for research laboratories and organizations to ensure the uniformity, consistency, reliability, reproducibility, quality, and integrity of products in development for human or animal health including pharmaceuticals.

GLP is also applicable to nonclinical studies carried out to evaluate the efficacy or safety of products under development including medicines for humans, animals, and the environment. Standards for laboratory safety – suitable gloves, eyewear, and clothing to handle lab items safely – are not the same as GLP data and operational quality system. Throughout the course of nonclinical and laboratory testing, the GLP principles seek to ensure and promote the safety, consistency, high quality, and dependability of chemicals. GLP covers more than just chemicals; it also covers things like medical equipment, food additives, food packaging, color additives, animal food additives, additional nonpharmaceutical products or materials, biological products, and even electronic products.

The planning, carrying out, monitoring, recording, archiving, and reporting of nonclinical health and environmental safety investigations are the focus of the GLP quality system. The GLP laws impose norms of laboratory behavior such as the employment of certified staff, instrumentation, and analytical methods.

They also create a requirement for correctly and comprehensively documenting such behavior. It is simple to recreate and audit the study after this has been done.

Together with the advantages listed above, GLP contributes to the veracity and traceability of data given, addressing the problem of nonreproducibility in many biopharmaceutical experiments. GLP aims to reduce negative pharmacological effects and enhance environmental and human health safety profiles. With the transparent and thorough documenting of laboratory activity, while allocating responsibilities at various stages of the experiment, GLP also contributes to improving accountability and the precision of results.

https://doi.org/10.1515/9783111191492-007

GLP is supported in the application of its concepts and elements by four pillars and five fundamental points or requirements. Figure 7.1 provides a simple illustration of these.

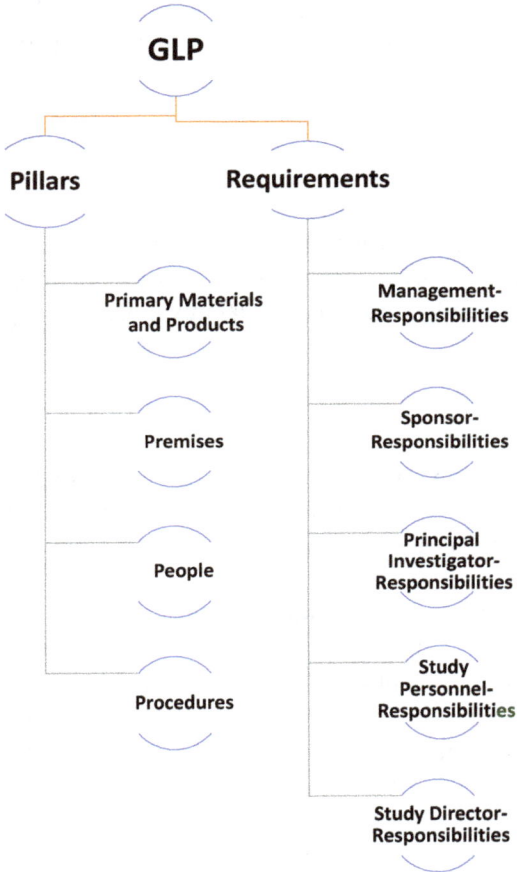

Figure 7.1: GLP pillars and requirements.

7.2 GLP Basic Elements

7.2.1 Personnel

The testing facility manager must select a study director prior to the start of the study, who will be in charge of the study's overall management and GLP compliance. It is also necessary to have a quality assurance unit (QAU) that is separate from or independent of the management or organization of the testing facility.

7.2.2 Facility and Equipment

To avoid interference and other disruptions that could jeopardize the study, the testing facility should allow for operation isolation. The control and test materials or samples must be received and stored in separate locations. Every piece of equipment used in the study must be maintained and calibrated on a regular basis. Equipment operators should have access to calibration and maintenance records that have been kept.

7.2.3 Characterization

Personnel conducting the study should be familiar with the following characteristics of each test and control material: identity, purity, composition, stability, date of receipt, expiry date, storage instructions, quantity received, and quantity used.

7.2.4 Study Plan or Protocol

The study plan or protocol serves as the primary guidance document for the study's execution. It describes how the study should be carried out and includes a general time schedule for the study and its various stages. It also includes the study's methodology and materials.

Before the study can begin, the protocol must be approved, reviewed, and discussed. The protocol is prepared by the study director, who then discusses its contents with personnel and other study staff. Following discussion, the protocol must be approved by the study director by affixing their dated signature. The protocol must be examined by the QAU to ascertain its compliance with GLP after being given the study director's approval. Personnel should now receive their own copy of the procedure as well as instructions on the responsibilities listed in it.

7.2.5 Standard Operating Procedures

The testing facility should have standard operating procedure (SOP) for each of its several departments, especially for common tasks. SOPs must be approved by the testing facility manager and any variations from SOPs need to be permitted by the study director.

7.2.6 Final Report

The study director, who drafts and approves the report, is ultimately in charge of the final report. Figure 7.2 provides an overview of the main components of the final report.

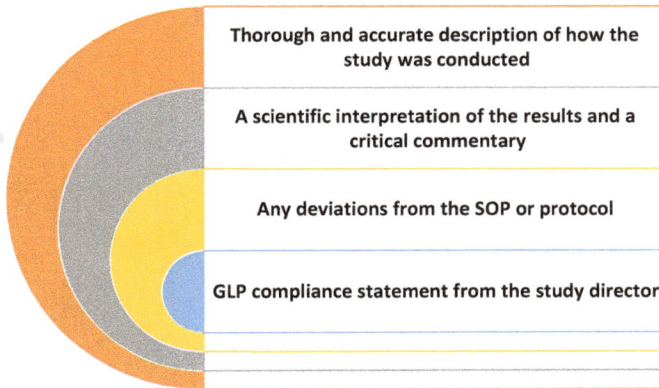

Figure 7.2: Main components of the final report.

7.2.7 Storage and Retention of Records

The study director is responsible for ensuring that all pertinent data is documented and incorporated in records that are safely stored during the study. At the conclusion of the study, these records and documents – including the protocol, the final report, and SOPs – will be archived. Anyone else wishing to access archived documents must first obtain authorization from the testing facility management.

A record must be kept each time it is accessed, removed from, or put back into the archives. It is also suggested that records in archives be indexed for streamlined retrieval.

7.3 GLP Principles

The principles of GLP serve to provide a sound approach to the administration of laboratory studies including conduct, reporting, and archiving and to encourage the generation of high-quality test results. The principles can be viewed as a collection of guidelines for making sure research are of high quality, reliable, and integrity, and that verifiable conclusions are reported together with data that can be traced back to its source. In order to ensure effective operational management of each study and to concentrate on those study execution-related activities that are particularly crucial for the reconstruction of the entire study, institutions are required by the principles to assign roles and responsibilities to staff.

As each of these factors is equally crucial for adhering to the GLP principles, it is not acceptable to apply GLP requirements only partially while maintaining your claim of GLP compliance. If a test facility has not adopted and complies with the entire set of GLP requirements, it cannot legitimately claim to comply. Figure 7.3 provides a synopsis of the 10 GLP principles.

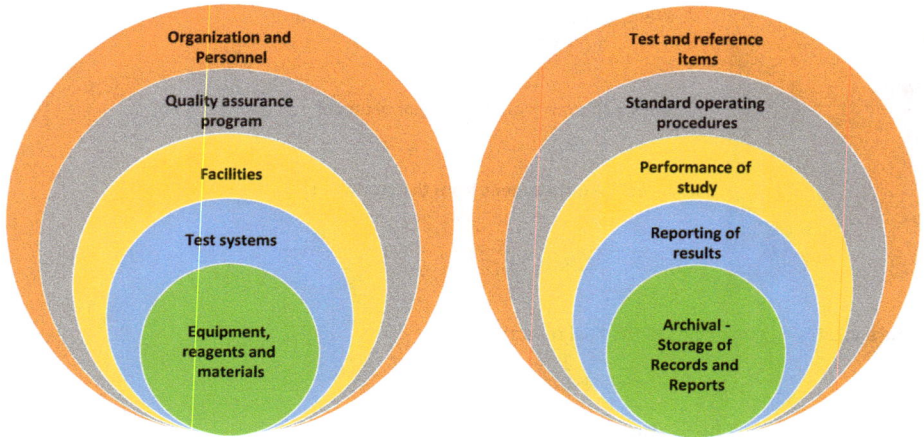

Figure 7.3: The 10 Good Laboratory Practice principles.

7.4 GLP Implementation

There are some basic prerequisites for this practice, which are enumerated as follows and must be met in order to adopt the GLP system in any organization:

- The laboratory has the necessary equipment, skilled staff, and approved documented procedures to carry out the investigation on schedule.
- The study is carried out in accordance with a preapproved study plan, and any deviations from the plan are recorded, analyzed, and authorized for their impact on the validity of the study's conclusions.
- The study's conduct will be overseen by a specific person who is qualified by a suitable set of education, training, and experience.
- All employees must be qualified to perform the tasks assigned to them and must be able to demonstrate their competence. They must also be free from any undue management, financial, commercial, or other pressures that might impair their judgment or prevent them from carrying out their responsibilities in a proper manner.

- The laboratory has a quality assurance procedure, which is overseen by employees who are not subject to any undue pressure that could skew their judgment or prevent them from doing their tasks properly. Direct access to the highest levels of management, where decisions on laboratory policy and resource allocation are made, is required for the individual in charge of the quality assurance program.
- All laboratory equipment used in studies is clearly appropriate for its use, all analytical testing procedures are appropriately validated, and all testing equipment is suitable for its use, performs in accordance with the necessary performance specifications, is maintained, and is calibrated on a regular basis in accordance with a written schedule. Each instrument shall be utilized only if it meets the set criteria.
- The lab takes the necessary procedures to avoid contamination and confusion between tests and reference materials. All discrepancies are extensively documented and looked into, and all analytical testing is recorded to show that testing was done according to the stated approved procedures.

7.5 Standard Operating Procedures (SOPs)

For successful GLP compliance, a complete collection of effective SOPs is required. The SOP system setup is frequently regarded as the most crucial and time-consuming compliance task. An SOP document provides guidance on how to conduct everyday tasks in order to boost productivity, standardize processes, and guarantee adherence to quality standards. With SOPs, all team members and organizational levels can communicate with one another. The management, department, function, or asset may order SOPs. Technical SOPs and management SOPs are two methods to conceptualize SOPs.

Technical SOPs provide instructions on how to carry out and finish tasks. These frequently take the shape of an inspection, a repeated work order, or a work order for preventative maintenance. Second, management SOPs describe the procedures for developing, revising, distributing, and monitoring all other SOPs. In essence, management SOPs describe the steps that must be taken to develop, record, and put into practice SOPs. In Figure 7.4, the SOP's life cycle is shown.

7.5.1 SOP Essentials

SOP essentials are always required in order to get the desired result. Those are:
- Keep an acute user perspective. The target user should always be considered while writing SOPs.
- Create a clear format. Process documentation must always be formatted clearly and use precise language.

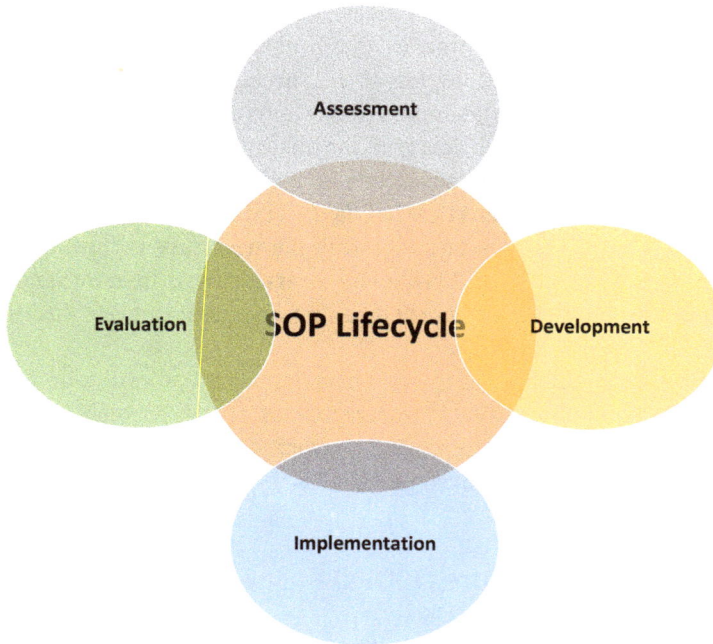

Figure 7.4: SOP life cycle.

- Think about the range.
- Watch the roles and effects.
- Request approval and power.
- Seek authority and approval while keeping an eye on roles and impacts.

7.5.2 SOPs Successful Implementation Requirements

One should keep the following in mind to ensure that the SOPs are implemented successfully:
- Consistent and fervent support from all levels of management, with a dedication to making SOPs a crucial component of the lab's structure and culture.
- Staff education and training are based on SOPs, ensuring that everyone follows the instructions exactly.
- A reliable SOP management solution to make sure that the most recent SOPs are readily available.

7.6 GLP QA Profession

The main purpose and role of the GLP quality assurance (QA) professional is to assure compliance with the GLP regulations within their departments and conduct audits of the facilities and monitor ongoing work in the facilities of various documents.

The GLP regulations require that QA conduct the following:

– Facility audits to ensure that they are suitable for purposes and that the necessary paperwork is in place to support processes.
– Process and study audits that involve watching staff members operate in the lab to ensure they are adhering to GLP and the appropriate procedures.
– Review the paper outlining the work to be done is included in the study plan.
– Review the research report to ensure that it accurately reflects the information gathered during the investigation and that it complies with GLP requirements.
– Examine the paperwork related to the validation of the new system together with audits of continuing maintenance and validation status.
– The role also requires proactive participation in changes to departmental working practices and policies and promotes professional and personal development by providing opportunities for training and consultation.

7.7 Tips to Achieve GLP

The following are some suggestions that could aid in putting the GLP into practice and achieving it in every particular workplace:

– Incorporate SOPs and follow them in the laboratory.
– Incorporate SOPs for calibration, maintenance, inspection, and testing. SOPs make it simpler to record difficult processes and lower variability between individuals and tests.
– Include all reagents' and solutions' names, opening dates, recommended storage conditions, and expiration dates.
– Ensure adequate dose selection and ascertain that all methods and tools are reliable.
– Link every piece of information to its sources and ensure that all findings are recorded not just those that support the theory. Make sure all documents are accessible for review. Any preset inclusion must be documented.
– The relevant project manager must sign and date each analytical report. Reports need to be preserved for at least five years. Documents should be systematically archived so that they are always accessible.
– Be careful you receive the requisite certification and credentials for performing the process, stay up to speed on all necessary procedures (and their pertinent SOPs), and follow excellent laboratory practices and procedures.

– In event of an emergency, be conversant with SOPs and equipped to follow them. Get familiar with the safety data sheet for the experiment's chemical inputs.

7.8 The Eight Examples of Good Laboratory Practice

GLP is managed by a quality control division or unit. Nonetheless, everyone on staff who participates in laboratory testing is ultimately accountable for it. Here are the top eight examples of GLP:

7.8.1 Audits and Inspections

To ensure that procedures are being followed, these must be carried out regularly. To determine whether the good practice system is operating properly and whether it requires any improvements, both internal and external audits are required. Research protocol, processes, and reporting are the main topics of audits.

7.8.2 Standard Operating Procedures (SOPs)

SOPs are the rules that laboratories and inspectors must abide by. All staff members should be able to easily access and understand the processes. A collection of forms or documents should be supplied in addition to the rules to record data. They ought to be complete, coherent, and consistent. The SOPs guarantee that standards are assessed and met. This confirms the effective production and trustworthy research findings.

7.8.3 Data Recording

Data collection covers the nature of the test, its execution, timing, and participants. To prevent errors, every data must be produced and documented accurately. Workers should never rely on recollection because incorrect information could be recorded and result in major problems. Traceability and accurate record-keeping are ensured by this procedure, whether they are kept on paper or digitally. It is recommended to utilize standard lab notebooks and to enter data into computers in the right format. Data should be readable by other staff members and be clear and simple.

7.8.4 Proper Equipment Use

Inaccurate data could be shown if incorrect or uncalibrated equipment is used. Before conducting any test, the equipment should be selected and calibrated. The designated tools and gear shall only be used for their intended purposes and be kept in good working order in accordance with a designated maintenance schedule. Data loss is avoided, and dependable data delivery is aided. It is advisable to keep records of equipment maintenance.

7.8.5 Examined Items

Test materials must always be marked and kept in the proper temperature, light/darkness, and humidity ranges. The sample should be labeled with the date, research information, batch number, and any other necessary data specified in the procedures.

7.8.6 Personnel

Only employees who are qualified to conduct specific testing methods should do so. Employees ought to be aware of their responsibilities and have job descriptions. Records of employee education, training, and development should be preserved. Typically, a study director who is in charge of the study's outcomes oversees pharmaceutical research studies. Their responsibility is to ensure that proper procedures are followed and to approve the study's results.

7.8.7 Proper Instruction

An important component of effective laboratory practice examples is training. Staff members who work in laboratories must comprehend the procedures to be followed and why they are crucial. This applies to training brand-new hires as well as retraining current employees to review testing, analysis, and reporting procedures as well as introduce brand-new ones.

7.8.8 Workplace Conditions

Every test should be run in a lab that has been set aside for that purpose. The environment must be at the proper temperature, there must be adequate room to move around, the test site must be clean before and after the test, and cross-contamination precautions must be followed. Employees should always dress appropriately for the lab and use the appropriate personal protective equipment.

7.9 Questions

7.9.1   What are the GLP four pillars?

7.9.2. What are the GLP five fundamental requirements?

7.9.3   List down the GLP basic elements.

7.9.4   What are the main components of the GLP final report?

7.9.5   List down the 10 GLP principles.

7.9.6   What are basic prerequisites for GLP Implementation?

7.9.7   Describe the SOP's life cycle.

7.9.8   What are the SOP essentials?

7.9.9   What is the main purpose and role of the GLP quality assurance professional?

7.9.10 What are the eight examples of GLP?

Chapter 8
Laboratory Information Management Systems (LIMS)

8.1 Introduction

A Laboratory Information Management System (LIMS) is a software-based laboratory and information management system that helps a modern laboratory run efficiently. It also improves lab productivity and efficiency by tracking data on samples, experiments, lab procedures, and equipment. It also allows technical employees and scientists to monitor samples and specimens at all stages of the analytical process such as when tests are run, test findings are reviewed, and control limits and quality control (QC) values are monitored.

LIMS is a type of computer software that is used in chemical laboratories to manage the following:
- Registration of samples
- Users of the laboratory
- Instruments
- Additional laboratory operations such as billing and workflow automation

Furthermore, LIMS capability has expanded beyond its primary aim of sample management to include assay data management, data mining, data analysis, and an electronic laboratory notebook. The LIMS system serves as a major reporting tool, allowing users to enter information about a test sample, such as the inspection number, the batch of material from which it was removed, and the date, time, and location. The LIMS system stores the sample's details as well as information about its location. The LIMS system may be updated as the sample progresses through the testing process, allowing users to see where each sample is at any time. A sample can be tracked by manually inputting the sample number into the system and manually entering the location or by employing barcodes. The LIMS can print a barcode label with the unique sample number when a sample is first entered into the system.

8.2 Basic Components of LIMS

Figure 8.1 shows the three main elements of an effective LIMS: tracking of samples, carrying out protocols, and storage organization.

https://doi.org/10.1515/9783111191492-008

Figure 8.1: Main elements of LIMS.

8.2.1 Tracking of Samples

The main purpose of LIMS is to follow a sample from the moment it enters a lab through testing and storage. This entails documenting all of the information related to the sample at the time of its initial accession including its ID, source, collection date, and quantitative details like concentration, volume, and particle quantity. Additional data is recorded when the lab sample moves through its workflow, and this information is likewise saved in the LIMS. This comprises test results, sample data that was derived, and metrics for time-based research. The LIMS records and tracks information particular to each lab sample as well as information on who has interacted with the sample and where it has been over the course of its existence.

8.2.2 Carrying Out Protocols

The second crucial function of laboratory information management system software is the standardization of a lab's operations and underlying protocols, methodologies, and stages. Regardless of who is processing the sample or conducting the test, LIMS ensures that each laboratory worker processes a sample in compliance with the specific instructions in a published SOP. A LIMS also facilitates standardization among laboratory workers by digitizing the processes in procedures and protocols. This ensures that when testing a sample, each member of the lab staff follows the proper steps in the proper order.

8.2.3 Storage Organization

Tracking a sample's whereabouts throughout its laboratory lifespan is the third essential component of a laboratory information management system. Starting with each individual lab sample, the LIMS keeps track of where the sample tube or vial is stored in a specific box. The system then maintains track of which rack and drawer each box is located in. Additionally, the system keeps track of the area and shelf that the freezer is in.

8.3 Advantages of LIMS

Scientific and commercial laboratories can enhance various QC and quality assurance procedures by using LIMS software to automate the entry of data from instruments and record, manage, and organize large collections of data for quick search and retrieval. The four main advantages of LIMS for each specific laboratory are enumerated in the list below.

8.3.1 Productivity and Efficiency

Productivity and efficiency are both increased by the LIMS-supported automation of the entire procedure. Less manual work is required at every stage of the procedure, which also lowers the possibility of human error. As a result, laboratories become more focused on results and have better access to their clients' knowledge and talent.

As a matter of fact, LIMS significantly improves and streamlines lab work by providing lab technicians with straightforward yet useful functions that allow them to operate more productively and efficiently and motivate them to offer their best rather than waste their time on needless and difficult operations.

8.3.2 Automation

LIMS frees lab employees from time-consuming chores by handling many of them automatically. For instance, when used to speed up the laboratory process, LIMS can automatically assign tasks to researchers or show the location of a sample's next destination. LIMS also has the capacity to streamline and automate laboratory inventory management. Additional LIMS capabilities include inventory management and equipment management. LIMS also provides a solution for the problem of instrument calibration and maintenance. It provides the opportunity to schedule necessary maintenance chores and lab instrument calibrations and keeps track of all related procedures.

8.3.3 Connectivity

Various laboratory devices and applications can be connected with LIMS systems for controlling certain activities and for data transmission. For the safety and security of the data, the integration can be controlled. To improve resource planning and data management effectiveness, they can also be integrated with accounting software and enterprise resource planning programs. Additionally, LIMS offers labs data integration that not only guarantees secure data management tailored precisely to the diverse demands of different labs but also greatly cuts down on the time required to receive information from a separate facility.

8.3.4 Cost-Cutting

LIMS lowers costs by boosting productivity through process automation, maximizing labor efficiency, reducing operational and regulatory risks, and addressing persistent bottlenecks like report and bill preparation. LIMS decreases the number of unnecessary resources used in various processes in addition to accelerating workflow and saving time.

The problem of document storage and the requirement for archiving facilities can be resolved by using LIMS to digitize formerly manual processes.

8.4 Disadvantages of LIMS

Although LIMS has numerous benefits and uses, it also has certain drawbacks, which are as follows:
- To make the switch to the software system, upskilling and employee training are crucial because LIMS calls for more highly qualified staff. The new system's adoption can incur considerable time and financial costs during the training process.
- Getting acclimated to a new system takes time for most people. However, a poorly designed system might lead to a lot of frustration owing to a subpar user interface (UI). The move to LIMS will be simpler and processes will run more smoothly with high-quality UI. While standardization is a key component of digitalization, personalization is also important. Operator productivity suffers when noncustomizable, rigid dashboards are used; thus, it is crucial to have flexible, dynamic dashboards that can host a variety of visuals and data.
- Scalability issues and subpar integration. A poor LIMS solution can make scaling and integration very difficult. Integration of the stiff software with other tools and apps would be difficult. The scaling-up of the solution to accommodate growing businesses can also become a significant problem due to extensive customization or bad design.

8.5 Functions of LIMS

8.5.1 Sample Receipt and Sample Login

LIMS is used to assign a job and create the bottle labels for all samples as soon as they arrive at the laboratory. All necessary sample information, including the tests to be run, must be entered during the sample login to complete the information. Even very long client sample identifiers can be captured by the sample login process.

8.5.2 Sample Scheduling

All sample scheduling and prioritization are managed by LIMS. This guarantees that all samples, even urgent and emergency samples, receive the proper attention from all lab departments. This covers crucial concerns like turnaround and hold times.

8.5.3 Data Acquisition

All analytical data collected from automated instruments are sent to LIMS electronically rather than manually, avoiding the inevitable transcription errors that can occur when manually entering large amounts of data. This includes all organics, metals, and even certain general chemistry tests.

8.5.4 Data Processing and Data Approval

All data enters the system through a strictly controlled "assembly line" of calculations, automated checks, and approval stages. The data must be reviewed by both the analyst and the supervisor. It should be noted that the system is not intended to replace the scientist; rather, it does all possible to make their job easier and more streamlined, resulting in a more thorough review process.

8.5.5 Quality Control

All relevant QC data, such as spikes, duplicates, and blanks, are entered and processed in the same way that live samples are, but with the addition of QC calculations and automatic QC limit testing against predetermined limits. This QC data is "linked" to all appropriate samples automatically. This important relationship enables the laboratory to integrate entire QC data in automated reports, electronic deliverables, and online date services.

8.5.6 Final Reporting

All sample summaries, result pages, and QC summaries are generated straight from LIMS when creating the final hard-copy report. Only backup data such as chromatograms and other types of raw data are independently gathered.

8.5.7 Quotes and Billing

LIMS is also in charge of maintaining all client quotations and invoicing data. All contract prices, discounts, turnaround surcharges, and even unique client billing requirements are included. This ensures that billing is accurate based on quotations and contracts.

8.5.8 Equipment Calibration and Maintenance

During inspections, the equipment used to extract and test the sample must be properly maintained and calibrated to introduce no errors in the test findings. The LIMS system should include maintenance data for testing equipment so that notifications can be produced to undertake routine preventive maintenance. Various instruments and equipment may require calibration after a particular number of usages. The calibration instructions can also be stored in LIMS systems, allowing a notification and a calibration instruction sheet to be delivered to the maintenance department or a specialized outside vendor.

8.5.9 Testing Methods

The LIMS system should store and manage the processes, procedures, and methodologies that may be utilized to perform the necessary tests at the facility. The system should provide a centralized repository for various approaches and be capable of selecting the appropriate way for the tests to be run.

8.6 LIMS Accuracy and Security

LIMS accuracy consists of the following:
– Error reduction
– Check the data at the time of input
– Improve QC
– Avoid late deadlines
– Avoid the disgrace of false results

The LIMS security, on the other hand, contains the following:
- Create secured access and maintain security by creating individual login accounts with tiered access capabilities that are managed by the lab's LIMS administrator
- Keep track of database revisions
- Keep the system safe while customizing it

8.7 Benefits of LIMS

The use of a LIMS system in laboratory benefits all areas of the business, not just the laboratory itself but also the laboratory's clients, whether internal or external. The pressures on a laboratory to increase performance in terms of sample turnaround and data quality are increasing, and a well-implemented LIMS system can help in these areas. Here are some of the primary benefits that a LIMS system can provide:
- Samples entered into the LIMS are assigned a unique number and barcode, and all actions done on the samples are documented including the person performing the action and the time the action was completed. This increase in traceability aids regulatory compliance, increases lab capacity, and reduces turnaround times.
- Capturing data directly from laboratory instruments as a feature of a LIMS improves data accuracy and speed of data entry, increasing compliance and enhancing lab performance. Reducing the number of retests also saves the laboratory time and money.
- A LIMS system offers several degrees of security and audit trails, allowing you to track activity and ensure responsibility. It also provides access to historical data through built-in searching and reporting features, which aid in data analysis and the utilization of this information to improve productivity.
- Many LIMS systems can handle consumable goods, keeping track of expiry dates and locations, and relating them to specific ways for controlling usage. The ability to detect reagent expiry and run reports against LIMS data to identify consumables nearing expiration can reduce the risk of compliance concerns and retesting.

8.8 Types of LIMS

LIMSs are essential in modern labs for everything from tracking samples and storing data to optimizing processes and scheduling equipment maintenance. There is no one-size-fits-all solution; therefore, facilities must invest in LIMS that meet their specific requirements. The following is a closer look at the different types of LIMS solutions that are customized to match the needs of various labs.

8.8.1 Clinical Evaluation and Diagnosis LIMS

Traceability is an important component of LIMS systems built for clinical analysis and diagnostic laboratories. Analysts must be able to track samples and create a complete chain of custody. LIMS of the most recent generation may also do sophisticated tasks like next-generation sequencing and quantitative polymerase chain reaction.

8.8.2 LIMS for the Environment

Environmental labs rely on samples, with some facilities processing hundreds of specimens per day. Labs that use gas chromatography-mass spectrometry (GC/MS) technologies to assess drinking water, for example, generate enormous data batches that can be handled by LIMS systems. It should also help to streamline procedures, cut expenses, and reduce administrative requirements.

8.8.3 LIMS for Academic Research

Academic Research LIMS software is designed to collect, analyze, store, and share data. Collaboration is at the forefront. The top LIMS systems provide tools that enable collaboration among university researchers simple and efficient.

8.8.4 Petrochemical and Chemical LIMS

Chemical and petrochemical companies deal with enormous amounts of data. These datasets can be processed by LIMS systems. Many industries rely on LIMS to optimize processes and comply with industry regulations and safety norms.

8.8.5 Food Industry LIMS

Strict health and safety standards apply to the food and beverage business. This entails following raw materials and finished commodities as they go through the supply chain. Food and beverage laboratories can use LIMS software to help them comply with food traceability standards. LIMS can also help to standardize the food manufacturing process and the finished product.

8.8.6 LIMS for Forensics

Forensic laboratories are subject to stringent norms and regulations. Accuracy and dependability are critical, and LIMS software assists labs in meeting the highest requirements. Watertight storage, handling, and chain of custody practices are also prioritized in forensic LIMS solutions.

8.9 Applications of LIMS

The LIMS can be utilized in a variety of lab settings including clinical and production labs. It is also used in scientific and commercial facilities to handle information and data. Some uses of LIMS are summarized in Figure 8.2.

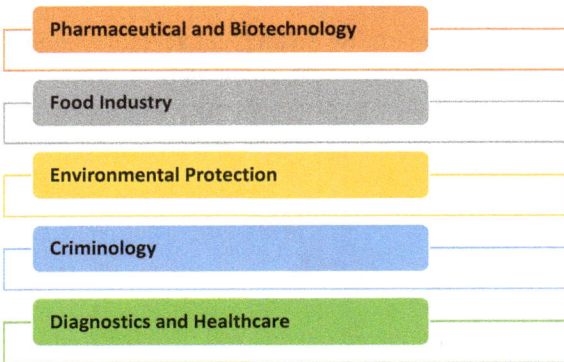

Pharmaceutical and Biotechnology

Food Industry

Environmental Protection

Criminology

Diagnostics and Healthcare

Figure 8.2: Applications of LIMS.

8.10 Implementation of LIMS

The benefits of deploying LIMS are undeniable. Proper planning and communication from the start may protect against common LIMS implementation challenges such as longer schedules, scope creep, unanticipated expenses, and poor user engagement. Before implementing a LIMS system, keep the following in mind.

8.10.1 Engage End Users

Involving the entire lab team, especially scientists, is crucial while preparing the lab for LIMS deployment. After all, they will be the end users. It is vital to understand their viewpoints and challenges. Their requests and ideas are crucial in developing the best possible solutions. Dispel any doubts and give them the opportunity to ask questions. Inform them about the possible benefits of utilizing LIMS and introduce them to new working conditions.

8.10.2 Define Your Workflows

It is best to describe your workflows before selecting a vendor. You can accomplish this by mapping your lab's workflows. To get visual, draw your process maps and suppliers, inputs, processes, outputs, and customers including decision points and parallel operations. Consult your map with end users and make any necessary changes.

8.10.3 Select the Vendor

Choose the best vendor using your process maps. By making conversations about your ideal LIMS more apparent, you can save a lot of time and money by presenting your process maps. LIMS companies' approaches to projects and deployments can vary greatly. That is why it is critical to select a reliable vendor who can satisfy your needs and specifications.

8.10.4 Design Your LIMS

To construct the best LIMS possible, your preferred vendor will arrange frequent meetings to gain a thorough grasp of your lab's requirements. It is critical to offer them relevant information and to be curious, ask numerous questions and involve lab technicians. While many laboratories rush to adopt a LIMS to manage their data, increase productivity, and fulfill business goals, it is preferable to invest some time and attention in building a LIMS and implementing a version that is ideal for them.

8.11 LIMS System Operation

LIMS software can include some of the following features:

8.11.1 Audit Management

An audit trail is a basic LIMS feature required for regulated laboratories, although it can be beneficial to any laboratory. Audit trails allow lab personnel to follow the audit path from start to finish, explicitly identifying the collaborator and the precise time of specific actions, or automating the audit process. LIMS can automate and control the audit process, ensuring that all regulations are met, and all stages are taken. It also detects abnormalities and reports them instantly to limit potential damage.

8.11.2 Barcode Handling

Handwritten labels can be completely replaced by barcodes. Barcodes provide various benefits to laboratory technicians handling samples by reducing the danger of unreadable information on the label and limiting the risk of human errors. Barcodes enable lab staff with more precise data input, tighter sample-instrument and sample-study linkages, and greater label space. Aside from the ability to design and print labels, a LIMS may also support a range of barcode readers.

8.11.3 Chain of Custody

The importance of the chain of custody varies depending on the laboratory. Tracking CoC is a crucial aspect of the operations of highly regulated laboratories. It is extremely significant in forensic labs and their evidence-gathering, retention, and disposal protocols. It is critical to retain and access information such as user ID or location ID to maintain successful CoC.

8.11.4 Compliance

LIMS can be used in laboratories across a wide range of sectors to organize and analyze data on samples, tests, and test results, among other things. Many of these businesses, as well as the laboratories inside them, are subject to stringent regulations. Compliance is critical since failure to comply with the regulations is likely to result in massive losses and falsifications.

8.11.5 Customer Relationship Management

Client demographic information is stored and managed by LIMS. It also streamlines and simplifies communication with linked clients. Furthermore, certain LIMS allows the client to request a sample from the LIMS. The sample container can then be registered and sent to the client as the next step. The sample can then be collected and delivered to the lab.

8.11.6 Document Management

In the typical laboratory, various types of data are gathered and kept. There are specifications, safety guidelines, regulatory information, and reports available.

The addition of the capacity to create, maintain, import, and export other sorts of data was a logical but also very helpful decision given that LIMS is already capable of managing sample and experiment data.

8.11.7 Instrument Calibration and Maintenance

It might be difficult to keep track of all the procedures involved in instrument calibration and maintenance, especially in a setting laboratory. LIMS provides the ability to schedule crucial lab instrument calibrations and maintenance procedures. Additionally, LIMS keeps thorough records of such actions and enables lab staff to continuously check the technical performance of lab equipment.

8.11.8 Inventory and Equipment Management

Administration of inventories takes a lot of time, effort, and concentration. However, things do not have to be this way. Inventory that can be measured, recorded, and controlled by a LIMS can be handled along with laboratory equipment.

8.11.9 Manual and Electronic Data Entry

Using the quick and dependable interfaces of LIMS, a lab technician can implement data with ease. This functionality facilitates data entry and enables it to be customized to the tastes and needs of a specific lab.

8.11.10 Method Management

The LIMS is the single location where all laboratory operations, procedures, and methodology are managed. It organizes and simplifies every lab procedure. In addition, LIMS instructs lab technicians on how to complete their current duties.

8.11.11 Personnel and Workload Management

The ability to schedule duties and events is provided by LIMS, which streamlines the workload management procedure.

Technicians can be given workloads, and maintenance plans can be made. Laboratory information management system can easily optimize all duties and streamline all lab operations.

8.11.12 Training Management

LIMS gives lab technicians the chance to schedule and plan training in addition to managing work schedules and storing employee data.

8.11.13 Quality Assurance and Control

The majority of lab procedures are under the control of LIMS, guaranteeing the highest level of quality. Furthermore, based on the type of lab, LIMS performs effective quality checks on samples or products. Additionally, it automates the logging of samples or products and rapidly extracts management information to guarantee complete process control.

8.11.14 Reports

LIMS features the ability for lab technicians to schedule and generate reports among many other features. To satisfy the requirements or rules of labs, they can be prepared in a particular, necessary format. Additionally, the report can be sent to specific people or groups after it has been created, enabling seamless and quick information delivery.

8.11.15 Time Tracking

The general ability of LIMS to monitor how much time an employee spends at work is referred to as the time-tracking function. Payroll-related uses for it are possible. Additionally, this feature can be used for projects and tasks that are more particular, and it aids in employee work evaluation programs by showing how quickly employees complete tasks and how much time they spend on each task.

8.11.16 Workflows

By automatically allocating duties to researchers or outlining the next location of a sample in the workflow, a LIMS can be used to streamline the laboratory workflow.

It can also be programmed to propose suggested instruments and equipment, based on preset rules and criteria, for specific procedures. LIMS avoids delays and keeps the lab workflow moving.

8.12 Types of LIMS Systems

The three most popular kinds of LIMS systems are cloud-based, standalone or web-based, and offline.

8.12.1 Offline Lab Software

From registration to reporting, actions are managed by offline lab software. The labs with the least amount of data handling and flow are best suitable for this software. However, it cannot support your lab business's higher data burden. Because of these drawbacks, offline systems are inappropriate for laboratories that are growing or have a lot of centers.

8.12.2 Standalone or Web-Based LIMS

This kind of lab management system can only be used online. It offers a one-time installation with no continuing support and goes by the name of a standalone LIMS. This online LIMS offers sufficient data storage but if the lab expands, it will be necessary to add newer versions.

8.12.3 Cloud-Based SaaS LIMS

It is considerably more sophisticated and adaptable than the other two types of lab software. It is flexible and scalable and offers more storage room. Because of this, when labs are growing, cloud-based LIMS software ensures there is no worry about data flow or operations handling. The best tools are now accessible in cloud computing and software-as-a-service (SaaS) methods for any laboratory analytical testing environment. The following benefits are available to lab managers who use cloud-based applications:

- No IT department requirement
- System validation worry-free
- No extensive upgrade projects requirement
- No time is spent in maintaining these lab automation systems

Without the burden of worrying about their software, lab managers and their teams can actually focus on their jobs. Put another way, they can focus on science not systems!

8.12.3.1 Advantages of Using a SaaS LIMS

The benefits of using SaaS LIMS are as follows:
- A SaaS LIMS can be completely set up and operational in as little as 30 days. In addition, execution and onboarding save a lot of time and money.
- A LIMS is anticipated to have validated, preconfigured workflows that are well-known to support laboratory operations. Best practices, such as barcodes and unique identifiers to monitor each sample through the entire testing process, are baked in because the system was designed with a knowledge of regulatory requirements across industries.
- An already validated system can reduce or eliminate the need for testing, validation, requirements analysis, and documentation procedures, saving up to 30–50% of the overall system cost.
- A well-known SaaS LIMS works efficiently by consistently storing data in the cloud and needing little local resources to run without lags or crashes. It is stored on web browsers, so it does not require solid or expensive PCs to function.
- The highest levels of security are provided by SaaS LIMSs, both within the program itself and at the infrastructure level. With a conventional LIMS, security is in the hands of the business installing the LIMS; they spend more money trying to safeguard their data from dangers like viruses and intrusions.
- SaaS LIMS has the benefit of not requiring a firewall or using a VPN, in contrast to LIMS hosted internally. Entering a SaaS LIMS only requires a login and password.

8.13 Questions

8.13.1 What are the basic components of LIMS?
8.13.2 What are the advantages of LIMS?
8.13.3 List down three disadvantages of LIMS.
8.13.4 What are the main functions of LIMS?

8.13.5   What is reporting in LIMS?
8.13.6   What are the stages of the LIMS workflow?
8.13.7   What are the LIMS software features?
8.13.8   List down five applications of LIMS.
8.13.9   What are the benefits of using LIMS?
8.13.10 What are the benefits of using Cloud-based SaaS LIMS?

Chapter 9
Working with Laboratory Glassware

A chemistry or chemical laboratory requires special glassware. Glassware items are needed to measure accurate volumes, to run chemical reactions, and to do extraction and purification. Borosilicate glass – which is used to make most glassware, including Pyrex and Kimax – is a common material. Some glassware is not truly made of glass; instead, it is an inert substance like Teflon.

A big part of working securely is being aware of hazards. Minor cuts are the most common consequence of laboratory glassware mishaps. Fire, chemical exposure, and risks from flying glass are among the more severe accidents. When performing laboratory tasks, put on safety goggles. Hazard awareness can spare you time and money in addition to reducing injury.

9.1 Glass Material Types

Silica is a mineral that is present in glass and sand. Three kinds of glass materials are typically used in laboratories: pure fused quartz (99% silica), borosilicate, and soda lime (soft). Working temperatures for soft, hard, and quartz glass can reach 110, 230, and 1,000 C, respectively.

9.2 Hot Glassware Items

The similar appearance of the hot and cool glass is an issue. Let hot glassware items always cool in conspicuously placed areas. Let the glassware cool for several minutes before handling. Avoid glass burns by using gloves and tongs but bear in mind that using them may make manipulating glassware item and objects uncomfortable.

9.3 Preventing Cuts

When hand-washing glassware, thick gloves are recommended. Glassware cuts can be very serious and are more prevalent than you might think. Without taking the necessary safety measures, inserting a glass stem into a rubber stopper can be hazardous. It is safer and simpler to complete the job if the glass is first lubricated. Grease from a lab works well, but deionized water is still preferable to nothing.

Grease or wet the plastic pipe before attaching it to the flask's side arm or condenser. Do not try to yank the tubing off of the glassware when removing it.

https://doi.org/10.1515/9783111191492-009

If it is feasible, lay the item down on the lab bench first. Near the glass's extremity, cut the tubing. Never cut close to your flesh. The tubing should then be cut lengthwise, and the material should fall off the glass tip.

9.4 Tubing Properties and Fittings

Understanding the properties of the tubing material can improve lab safety by lowering the possibility of accidents happening. Plastic tubing is most commonly connected using a bolt closure, plastic barb fitting, and synthetic "rubber" gasket. Many manufacturers offer "Quick-connect" fittings as well. Glass is semi-permanently embedded with the connector's sole component. Ground-glass joints are the most commonly used way of joining laboratory glass apparatus. These are typically cylindrical or curvy in shape.

9.5 Pressure and Vacuum

When using glassware under pressure or vacuum, additional precautions should be taken. The most prevalent flaw that causes weakness and breakage is surface scratches. Before applying pressure or vacuum, examine glassware for minor flaws.

Mechanically pressurized or vacuum pump systems should be used in a fume hood with the sash down if at all feasible. Reducing the possibility of glass breakage, pressure relief, and vacuum-relief devices can reduce hazards and improve research results.

Make sure the tubing is securely clamped to the apparatus if an experiment is meant to generate positive system pressure or has the potential to do so. When working with vacuum systems outside of a fume hood, think about using epoxy-coated apparatus or taping the vessel to help confine glass in the event of a failure. When feasible, cover the benchtop. Keep in mind that round vessels have a greater capacity to endure pressure or vacuum than flat-sided vessels with an equivalent design.

9.6 Glassware Repair

By annealing, star fractures and other minor flaws can be "repaired." Glass is heated to a specific temperature and then slowly cooled as part of the annealing process.

The applied temperature rises as the glass becomes "harder." Glass stress is a sneakier threat associated with glasses. When heated unevenly above its strain point, glass can become strained. Quartz glass is challenging to stress, but borosilicate glass, which has a strain point of 510 C, is comparatively simple. Furthermore, thick glass experiences the most extreme thermal strain.

By removing the tension, annealing improves the reliability and safety of glassware. Glassware that has chips may be less durable and a risk for harm. Make sure to empty and sanitize each piece of glassware before sending it in for repair. Glassware should be rinsed with water after using acetone or another flammable fluid and then allowed to dry.

9.7 Types of Glassware in the Laboratory

Depending on their intended use, a variety of glassware items can be made in a laboratory from various kinds of glass. For instance, amber glass is ideal for storing liquids because of its darkened color, and quartz glass transparency is suitable for normal laboratory practices, while heavy-wall glass has been specifically strengthened for use in pressurized experiments. Figure 9.1 displays a selection of the glassware items commonly used in labs.

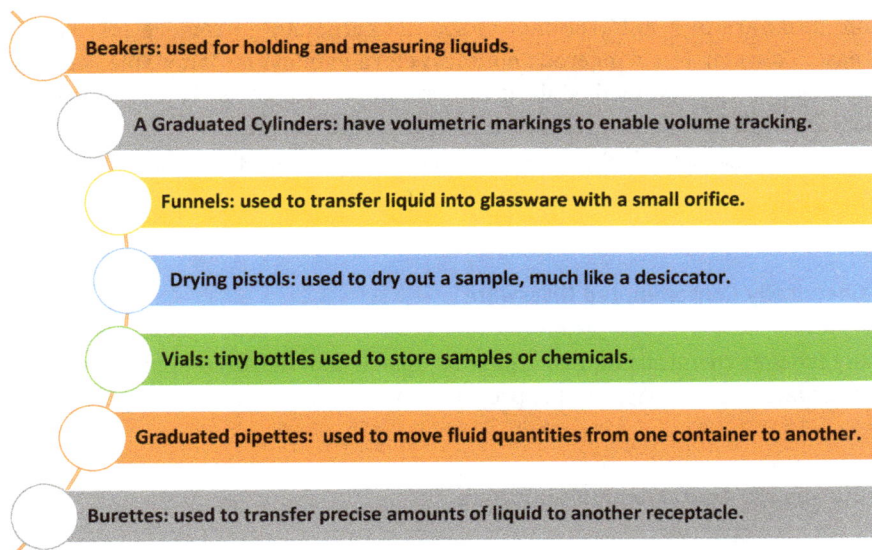

Beakers: used for holding and measuring liquids.

A Graduated Cylinders: have volumetric markings to enable volume tracking.

Funnels: used to transfer liquid into glassware with a small orifice.

Drying pistols: used to dry out a sample, much like a desiccator.

Vials: tiny bottles used to store samples or chemicals.

Graduated pipettes: used to move fluid quantities from one container to another.

Burettes: used to transfer precise amounts of liquid to another receptacle.

Figure 9.1: Display of a selection of glassware items and their use in labs.

9.8 Glassware Cleaning

Glassware used in laboratories needs to occasionally be cleaned, and this can be done in a variety of ways. To get rid of grease and loosen most contaminations, soak glassware in a detergent solution. The contaminants are then cleaned with a brush or scouring pad to get rid of any debris that will not come off during rinsing. Instead of cleaning, strong glassware may be able to tolerate sonication. Glassware is typically triple-rinsed after cleaning is complete before being hung upside down on drying racks. The following points should be kept in mind when washing glassware:

- After use, wash used glassware items as soon as you can. Glassware should be soaked in hot, soapy water until it can be cleaned. Avoid touching the sides of the sink or the water faucet with the glasses. This is how most breaks happen.
- Put on the proper gloves to protect your hands from the cleaning agents such as detergent, acid, and solvent as well as the residue left behind by dangerous compounds.
- After washing, completely rinse out any soap, detergent, or other cleaning supplies used, then let the item air dry.
- Use specialized cleaning chemicals or detergents like Decon 90 or Nochromix mixed with sulfuric acid to clean glassware that has become clouded, filthy, or includes coagulated organic matter.
- Avoid using brushes whose spine is visible due to wear and tear. When the brush touches the glass, it could cause severe scratches.
- Boil glassware in a mild solution of sodium carbonate to remove lubricating grease. Consider less dangerous options before employing flammable solvents like acetone. The best method for getting rid of silicone grease is to immerse the glassware in warm decahydronaphthalene (Decalin) for 2 h.
- Use racks made specifically for drying glassware to store the items. To prevent breaking, make sure the different pieces do not touch one another.

9.8.1 Chemically Contaminated Glassware

Pour any leftover chemicals into a garbage receptacle that has been appropriately labeled. Use warm water, soap, and the proper cleaning brush to scrub your glassware.

9.8.2 Biologically Contaminated Glassware

Glassware should be autoclaved using cycle parameters that are suitable for the biological agent being used. Empty the autoclaved biological material into the proper waste container after the glassware has cooled to room temperature. Use warm water, soap, and the proper cleaning brush to scrub your glassware.

9.9 Laboratory Glassware Disposal Guidelines

A prescribed protocol must be followed for the correct disposal of laboratory glassware. All laboratories that use glassware must have bins for broken glass intended for noncontaminated broken glass disposal. For their own use, laboratories can build their own broken glass containers by procuring a cardboard box, erasing any markings on the outside, writing "Broken Glass," and covering the interior with a trash bag.

9.9.1 Broken Glassware Disposal

Make sure that any shattered or broken glass is free of chemical and biological risks before disposing of it. Glass should be placed in a nonpuncturing container, which should be marked "broken glass" and sealed. Because they are simple to assemble and safe and convenient for disposing of broken glassware, disposal cartons offer a secure location to discard broken glassware. The following guidelines should be followed when disposing of shattered glass:
- Glass that has been broken must be handled with extreme caution. Always put on the necessary personal protection equipment such as protective eyewear, gloves, fully covered limbs, and close-toed shoes before handling broken glass.
- Only breakable glassware should be disposed of in broken glass containers. Broken glass should never be used with other materials.
- If the shattered glassware was not contaminated, it should be disposed of in the broken glass container; but, if it included chemicals or hazardous waste, it should be carefully gathered in a puncture-proof container and disposed of as hazardous garbage.

9.9.2 Broken Glass Container Disposal

The broken glass containers must be disposed of by the individual who made the broken glassware. Broken glass containers will not be removed or disposed of by custodial staff. It is required to keep broken glass containers inside the lab until they are thrown away. Glass containers that are cracked or broken should not be kept in hallways with regular trash cans. Do not overfill the broken glass container; fill it no more than 3/4 full. Gently tape the inside of the box's bag and close when you are ready to discard the broken glass container and seal the box.

Take the broken glass containers out of the lab and put them in the dumpsters for solid garbage, which is often found in the loading docks. When disposing of shattered glass containers, always wear personal protective equipment. When putting glass containers in metal dumpsters, use the utmost caution. Glass can hurt people if it breaks when it comes in touch with a harsh metal surface. When throwing away glass containers, always wear eye protection.

9.10 Questions

9.10.1   What are the glass types?

9.10.2   How is hot glassware handled?

9.10.3   Briefly explain how to use glassware under pressure or vacuum.

9.10.4   Briefly explain how glass is repaired.

9.10.5   What are the drying pistols used for?

9.10.6   How is the broken glass disposed of?

9.10.7   Give three examples for the common glassware items used in the laboratory.

9.10.8   How the biologically contaminated glassware items can be cleaned?

9.10.9   Briefly explain how generally glassware items are cleaned.

9.10.10 What are the guidelines that should be followed when disposing of shattered glass?

Chapter 10
Laboratory Emergency First Aid

10.1 Lab-Related Emergencies

10.1.1 Inhalation of Gases and Vapors

Asphyxiation from inhaled gases and vapors can occur due to a lack of air, which in turn results in other breathing issues. In either situation, it is essential to remove the affected person to stop further exposure to harmful gases and to get them to a place with clean air.

The safety of the environment in the affected person's area should be evaluated before trying the rescue. Cardiopulmonary resuscitation (CPR) should be administered to a person who is not breathing well. In some cases, it may be necessary to give high-flow oxygen, which calls for an ambulance crew.

10.1.2 Skin and Eye Exposure to Chemicals

In almost all circumstances, it is recommended to thoroughly rinse the skin and eyes for at least 15 min. The only deviations from this rule are skin contact with solid chemicals that react with water. In these circumstances, it is necessary to first scrape the majority of the chemical off the skin. For eye exposure and significant skin exposure, immediate transport to an emergency room or a physician is advised.

10.1.3 Burns

Burns caused by chemicals, radiation, electricity, or heat need to be treated carefully and promptly. Naturally, the heat source needs to be eliminated first. Second, if it is simple to do so, take off any clothing and jewelry in the affected region. Do not pull on clothing that has burned or melted next to flesh if it is still attached. Tap water can be used to quickly continue cooling the affected area if there are no blisters or closed blisters present due to burns. A burn center or emergency room should receive the injured individual. Use only sterile water on exposed blisters because nonsterile water can lead to infections. Never use any lotion or ointment on a burn.

https://doi.org/10.1515/9783111191492-010

10.1.4 Electric Shock

Make sure the affected individual is not still in contact with energized electricity or that the source of the electricity has been completely turned off before continuing. Check the person's breathing and heartbeat, ask for more help, have someone dial the local emergency number, and administer CPR if necessary.

10.1.5 Cuts or Open Wounds

The reaction is determined by the extent of the cut. It goes without saying that a minor cut can be cleaned and dressed. A sizable wound that is heavily bleeding poses a significant danger. Apply direct pressure to the wound, ideally with a sterile dressing, to stop heavy bleeding. Quick action takes precedence over a slow search for a sterile dressing. Avoid removing an impaled item because doing so will probably make the bleeding worse. To keep the object in position and reduce bleeding, wrap it in a dressing. Shock, which is a shortage of oxygen perfusion at the cellular level, can result from excessive blood loss. Until help arrives, keep the sufferer warm and elevate their feet.

10.1.6 Exposure to Biological Agents

Any virus or bacteria exposure necessitates transport to an emergency center for evaluation. Make sure that the rescuer does not contract the same pathogen. There is no "first aid" process beyond separating the affected person and the pathogen. Provide ambulance staff with as many details as possible about the exposure such as the agent, duration, and type of exposure. When this happens, the receiving hospital must take extra precautions because they must consider the affected person as "contaminated."

10.1.7 Exposure to Radiation

After exposure to alpha, beta, X-ray, or gamma radiation, there is little that can be done in the way of "first aid," much like with biological diseases.

Shut down or remove the radiation source or otherwise separate the patient from the radiation source, with proper consideration for your own safety. As much information as you can about the exposure should be provided to the emergency staff.

10.2 Essentials of Laboratory First Aid

Due to the presence of hazardous chemicals, corrosive liquids, poisonous chemicals, and toxic fumes, laboratories can be dangerous places. Nobody should be alarmed by this, though, as the risks can be reduced by implementing workplace safety procedures.

As the name suggests, first aid is a collection of procedures that can lessen the effects of chemical exposure or an injury before professional medical assistance can be given. First-aid procedures should be known to everyone working in the lab as they reduce the severity of damage and save precious time in the event of a lab accident. Each lab should have personnel who have undergone first-aid training.

10.2.1 Standard Operating Procedure on First Aid

First aid should be covered in each laboratory's standard operating procedure, which should be clearly displayed in the space. It should have directions on what to do in the event of lab mishaps and crises. All or some employees in the labs should be familiar with the seven fundamental first-aid steps. Figure 10.1 provides a summary of them.

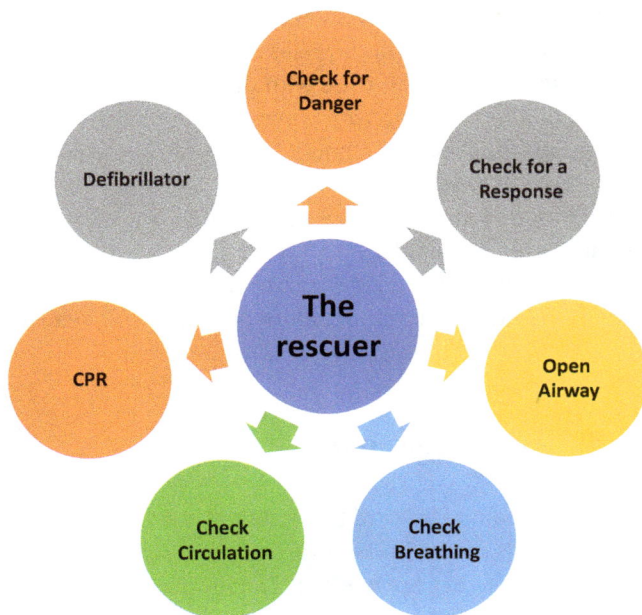

Figure 10.1: Seven fundamental first-aid steps.

10.2.2 First-Aid Kit

A first-aid package should be kept in a separate cabinet or box and should include all necessary medications. A person should be assigned the responsibility of keeping a record of the contents and discarding expired medications and replacing them with new ones. The following are the recommended minimum components of a first-aid kit:

- Individually wrapped plasters
- Sterile eye pads
- Triangular bandages
- Medium sterile wound dressings
- Large sterile wound dressings
- Disposable gloves
- Safety pins
- Antiseptic lotions
- Bandages and sterilized cotton

10.2.3 Specific First-Aid Tips

- Burns: Soak the affected area in running water for at least 10–15 min. Keep the wound open and do not apply any ointment until professional medical attention is available. In the event of severe acid burns, rinse with dilute ammonia (1–2%) or sodium bicarbonate solution after washing with water. Never add acid or alkali to the epidermis to neutralize the corrosive liquid. Things can become even more complex due to the heat of the reaction.
- Eyes: If corrosive liquid seeps into your eyes, rinse them completely with fresh water using an eye fountain or eye wash bottle.
- Poisons: Dilute the stomach contents by having the casualty consume one to two glasses of water and attempt to induce vomiting before poison-specific expert medical help can be given.

10.3 Preventing Lab Accidents at Workplace

Working in a lab necessitates complete knowledge of the nature of the work and the details of the job at hand. Professionals must also be aware of any hazards or risks associated with the undertaking on which they are working. Familiarize yourself with the instructions and specifics before beginning a new task or procedure.

10.3.1 Preparation

Ensure you are adequately prepared for the job before you begin. This entails:
- checking your tools and containers for any cracks or chips that might indicate vulnerability;
- checking for harm to electrical devices such as frayed cords or exposed wiring;
- be aware of the locations of the eyewash facilities;
- make sure your first aid kit is correctly stocked by giving it an once-over; and
- remove any clutter and look for potential tripping dangers.

Follow these guidelines for clothing to ensure that you are dressed for protection and success:
- Wear long sleeves that fit correctly, long pants, or long skirts to protect your arms and legs.
- Put on a pair of closed-toed, solid-material shoes with low heels.
- Tie back any long tresses.
- Take off any jewelry, particularly any rings that might catch on the gloves.
- When working with chemicals, use appropriate gloves and goggles. If you are pouring larger amounts, use a face shield to protect your neck and ears.

10.3.2 Prevent Contamination

You must take the following steps to avoid cross-contamination when working with substances in a lab:
- Never consume food or drink in a lab. For safety reasons, keep lotions and other makeup out of the lab.
- Remove gloves, dispose of them according to your arrangement, and wash your hands right away.
- Handle sharp objects, such as needles and syringes, carefully and discard them in the proper receptacle.
- Properly contain materials, particularly when taking them outside of the lab.
- Always put on clean PPE, such as a lab coat, goggles, gloves, and, if necessary, a respirator.

10.3.3 Care with Chemicals

- To prevent fumes, use a chemical fume hood and work in a well-ventilated location.
- Never taste, breathe, or inhale chemicals.

- Never pour chemicals down the toilet; instead, be aware of their proper storage location.
- When not in use, make sure containers are secure and lids are on securely.

10.4 Laboratory Must Have Things

- Eye wash units are used to rapidly flush an eye that has been contaminated by a chemical or biological substances. These units can be either bottled or plumbed in. All labs must have eye protection on, which lessens the chance of this happening. However, you must become familiar with their sites and operating procedures.
- Spill kits for chemical and biological spills are provided for containment, neutralization, and cleanup. While the contents of the two kits may be identical and contain either granules, or absorbent socks, one must be careful to use the right spill kit. If used on a chemical spill, specific biological spill control granules will react and produce an exothermic reaction and vapor release.

10.5 Questions

10.5.1   Give examples of lab-related emergencies.
10.5.2   What kind of first-aid procedure should be followed in case of exposure to alpha, beta, X-ray, or gamma radiation?
10.5.3   What are the seven fundamental first-aid steps?
10.5.4   What are the recommended minimum components of the first-aid kit?
10.5.5   What is the first-aid procedure in case of burns?
10.5.6   What are the uses of eye wash units?
10.5.7   What is the purpose of the spill kits for chemical and biological spills?
10.5.8   How can lab technicians prevent direct contact with fumes?
10.5.9   What kind of emergency procedure needs to be followed in case of any virus or bacteria exposure?
10.5.10 To whom should CPR be administered?

Chapter 11
Solvents Commonly Used by Chemical Technicians

11.1 Common Solvents for Chromatography

Many organic liquids are used in chromatography in general. In column chromatography, mixtures of less-polar and more-polar solvent systems are used. Most columns will benefit from hexane–ethyl acetate combinations. Pentane with diethyl ether or methylene chloride is another useful mixture, particularly for volatile products. Water, methanol, and acetonitrile are the most frequently used solvents in high-performance liquid chromatography.

11.2 Common Solvents for Extraction

Organic extracting solvents with moderate polarities, such as hexanes, toluene, dichloromethane, and diethyl ether, are commonly used. Table 11.1 depicts selected solvents and their characteristics.

Table 11.1: Selected extraction solvents and their characteristics.

Solvent name	Density (g/mL) at 25° C	Solubility in water (g/100 mL)
Acetone	0.784	Miscible
Chloroform	1.479	0.795
Diethyl ether	0.706	7.5
Ethanol	0.785	Miscible
Ethyl acetate	0.895	8.7
Hexane	0.659	0.014
Toluene	0.865	0.05
Methanol	0.787	Miscible
Dichloromethane	1.325	1.6

11.3 Common Solvents for Chemical Reactions

A solvent can perform a variety of roles in a reaction. It dissolves the reactants in the first place. The reactants might be solids without the solvent, or if they are liquids, they might be too thick for molecules to move around very rapidly; they might be more like oils. Since solvents typically have low boiling points, they easily vaporize or can be eliminated through a variety of straightforward procedures known as distillation, leaving the dissolved substance in its wake. Figure 11.1 provides a summary of the solvents that are most frequently used.

https://doi.org/10.1515/9783111191492-011

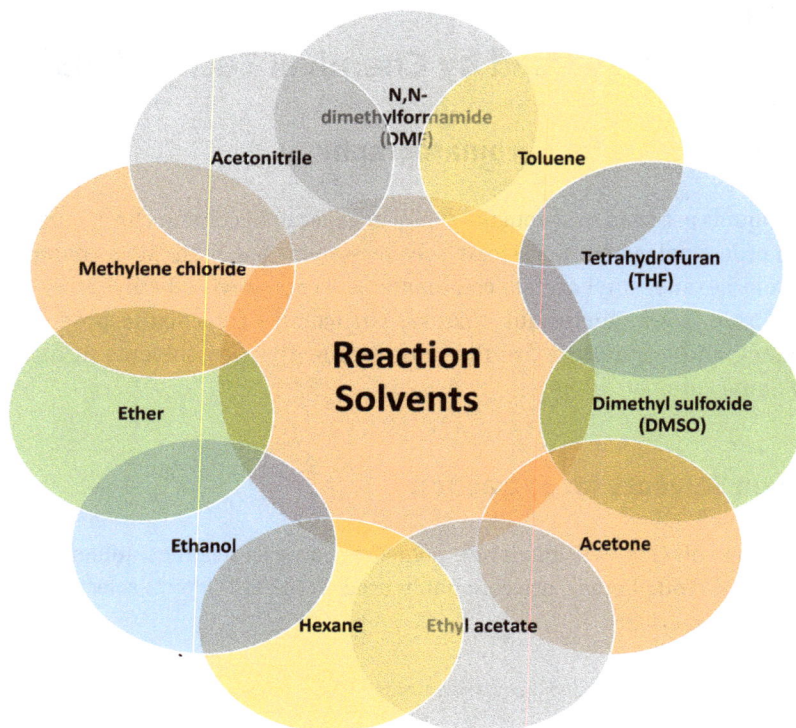

Figure 11.1: Commonly used solvents in chemical reactions.

11.4 Common Solvents for Characterization

Chloroform-D, acetone-d_6, benzene-d_6, deuterium oxide (D_2O), DMSO-d_6, ethanol-d_6, and methanol-d_4 are typical solvents used in the characterization of NMR samples.

The solvents carbon tetrachloride (CCl_4), carbon disulfide (CS_2), and chloroform ($CHCl_3$) are frequently used in IR solution cells; they are typically clear in the crucial absorption region of the spectrum. It is therefore possible to prevent substantial solvent interference by allowing a small path-length difference of the solvent between the background and the sample. Chloroform, methylene chloride, acetonitrile, and acetone are used for polar substance samples.

Water, methanol, acetonitrile, chloroform, tetrahydrofuran, and dichloromethane are typical solvents that are widely accepted and used for mass spectrometry analysis.

The most frequently used solvents in UV spectroscopy are those that are not absorbed in the area being examined. The most common solvents are listed in Table 11.2.

Table 11.2: Common solvents used in recording UV spectra.

Solvent	Wavelength (nm)
Water	191
Diethyl ether	218
Methanol	210
Ethanol	210
Chloroform	245
Carbon tetrachloride	265
Benzene	280
Tetrahydrofuran	220
Toluene	286
Pentane	210
Hexane	210
Ethyl acetate	255
Dichloromethane	235
Acetone	330
Acetonitrile	190
Cyclohexane	210
1,2-Dichloroethane	226
Dimethyl sulfoxide	265
N,N-Dimethylformamide	270

11.5 Questions

11.5.1   Ether and water have different densities. Water has a density of 1.000 g/mL and a specific gravity of 1.000. Specific gravity is the ratio of a compound's density to that of water. It is no surprise that water has a specific gravity of 1.000. If ether has a specific gravity of 0.977, which compound is heavier? Which compound will float to the top and which will sink to the bottom?

11.5.2   Methanol has a specific gravity of 0.980. When methanol and water are mixed, which compound should float to the top?

11.5.3   State whether the following organic solvents would form the top or bottom layer when mixed with water ($d = 1.0$ g/mL):
   –   Dichloromethane ($d = 1.33$ g/mL)
   –   Hexane ($d = 0.66$ g/mL)
   –   Toluene ($d = 0.87$ g/mL)
   –   Ethyl acetate ($d = 0.902$ g/mL)
   –   Chloroform ($d = 1.489$ g/mL)

11.5.4   Why is 95% ethanol generally not suitable as a solvent for UV-Vis measurements?

11.5.5   Why is acetone not a good solvent for UV spectroscopy?

11.5.6   What are the most commonly used solvents for chromatography?

11.5.7   Give four examples of the common solvents used in recording UV spectra.

11.5.8   What are the typical solvents used in the characterization of NMR samples?

11.5.9   Give five examples of the solvents that are most frequently used in organic reactions.

11.5.10 Is acetone miscible with water?

Essential Terms

Atomic absorption spectroscopy: A spectroanalytical technique that exploits the optical radiation (light) absorption of free atoms in the gaseous state to quantify chemical elements.

Broad-mindedness or open-mindedness: Accepting new ideas and being open to suggestions.

Chemical laboratory technicians: Technicians who assist chemists and chemical engineers in the research, testing, chemical processes, and product development.

Chemical plant operator: A technician or engineer responsible for running plant equipment safely and efficiently.

Chemist: A person who studies chemistry or works with chemicals or studies their reactions.

Composite sample: Grab samples that are taken multiple times at the exact location.

Corrosive chemicals: These are those that have the ability to directly corrode metal or burn, irritate, or damage living tissue.

Decahydronaphthalene (decalin): A widely used industrial solvent.

Decon 90: A cleaning solution used to clean contaminated glassware.

Documentation skills: The ability to maintain accurate records at work.

First-aid kit: A first-aid package that should be kept in a separate cabinet or box and should include all necessary medications.

Good Laboratory Practice: GLP is a quality management system for research laboratories and organizations to ensure the uniformity, consistency, reliability, reproducibility, quality, and integrity of products in development for human or animal health including pharmaceuticals.

Grab sample: A single sample taken at a specific time and location or a measurement taken at a specific time or over the shortest time period.

HAZWOPER: This categorizes employees who may contact chemicals into five emergency response levels and call for different training at each level.

Knowledge or cognitive skills: These are general mental skills involving reasoning, problem-solving, planning, and experience-based learning.

Laboratory Information Management System: A program that manages samples and associated data and combines tools and automate procedures.

Nochromix: A glassware cleanser made of a white, crystalline, inorganic oxidizer that is supplied in premeasured, hermetically sealed pouches to get rid of silicones and tough stains.

https://doi.org/10.1515/9783111191492-012

Researcher: A person who conducts academic or scientific research to discover new information or gain new understanding.

SARA III: This promotes effective community reaction and planning in the event of chemical release.

SARA 302 and 303: Emergency planning.

SARA 304: Emergency release notification.

SARA 311 and 312: Hazardous chemical inventory.

SARA 313: Toxic chemical release inventory.

Scientist: A person who has studied or acquired knowledge in one or more of the natural or physical sciences.

Skill: The ability to apply knowledge effectively and quickly in the performance of a task.

SOP document: A document that provides guidance on how to conduct everyday tasks in order to boost productivity, standardize processes, and guarantee adherence to quality standards.

X-ray diffraction: A popular technique for identifying the crystallinity and structure of solid samples.

Abbreviations

AAS	Atomic absorption spectroscopy
BOD	Biochemical oxygen demand
BTEX	Benzene, toluene, ethylbenzene, and xylene
CCl_4	Carbon tetrachloride
$CHCl_3$	Chloroform
CoC	Chain of custody
COD	Chemical oxygen demand
CPR	Cardiopulmonary resuscitation
CS_2	Carbon disulfide
D_2O	Deuterium oxide
DRO	Diesel range organics
EDX or EDS	Energy dispersive X-ray spectroscopy
EPA	Environmental Protection Agency
ERP	Enterprise resource planning
FTIR	Fourier-transform infrared spectrometers
GC/MS	Gas chromatography-mass spectrometry
GLP	Good laboratory practice
GMP	Good manufacturing practice
HAZCOM or HCS	Hazard Communications Standard
HAZWOPER	Hazardous Waste Operations and Emergency Response Standard
HPLC	High-performance liquid chromatography
IR	Infrared spectroscopy
KPI	Key performance indicator
LIMS	Laboratory Information Management System
MS	Mass spectrometer
MSDS	Material safety data sheet
NMR	Nuclear magnetic resonance
OSHA	Occupational Safety and Health Administration
PPE	Personal protective equipment
SaaS	Software-as-a-service
SARA III	Title III of the Superfund Amendments and Reauthorization Act
SDS	Safety data sheet
SEM	Scanning electron microscope
SOP	Standard operating procedures
SQC	Statistical quality control
QA	Quality assurance
QAU	Quality assurance unit
QC	Quality control
qPCR	Quantitative polymerase chain reaction
TCLP	Toxic characteristic leaching procedure
TEM	Transmission electron microscope
TPS	Toyota Production System
UI	User interface
UV-Vis	Ultraviolet-visible spectrophotometry
VPN	Virtual private network
XRF	X-ray fluorescence
XRD	X-ray diffraction

https://doi.org/10.1515/9783111191492-013

Resources and Further Readings

[1] Ballinger, J.T., Shugar, G.J. Chemical Technician's Ready Reference Handbook, 5th edition, New York: McGraw-Hill, 2011. ISBN 0071745920.

[2] Kenkel, J. Analytical Chemistry for Technicians, 4th edition, London: CRC Press, Taylor & Francis, 2013. ISBN 9781439881057.

[3] Iyengar, G.V., Subramanian, K.S., Woittiez, J.R.W. Element Analysis of Biological Samples: Principle and Practices, 1st edition, New York: CRC Press, 1998. ISBN 9780849354243.

[4] Ebdon, L., Fisher, A.S., Hill, S.J., Evans, E.H. (eds.). An Introduction to Analytical Atomic Spectrometry, New York: Wiley, 1998. ISBN 9780471974185.

[5] Welz, B., Sperling, M. Atomic Absorption Spectrometry, 3rd edition, New York: Wiley-VCH, 1999. ISBN 9783527285716.

[6] Harris, D.C. Quantitative Chemical Analysis, 7th edition, USA: W. H. Freeman, 2006. ISBN 9780716770411.

[7] Jenkins, R. X-Ray Fluorescence Spectrometry, 2nd edition, New York: Wiley, 1999. ISBN 9780471299424.

[8] Gill, R. (ed.). Modern Analytical Geochemistry: An Introduction to Quantitative Chemical Analysis for Earth, Environmental and Material Scientists, 1st edition, New York: Routledge, 2014. ISBN 9781138140820.

[9] Bruno, T.J., Svoronos, P.D. CRC Handbook of Basic Tables for Chemical Analysis, Boca Raton, FL, USA: Taylor & Francis, 2010. ISBN 9781420080421.

Internet Resources

– Collaboration Skills. Accessed on 8-2-2023. https://ceelegaltech.com/collaboration-collaborative-teamwork-for-futureprooflegal-014/.

– Pharmaceutical Lab Technician. Accessed on 9-2-2023. https://www.careermatch.com/job-prep/career-insights/profiles/pharmaceutical-lab-technician/.

– Medical Laboratory Technician. Accessed on 10-2-2023. https://www.careerexplorer.com/careers/medical-laboratory-technician/#what-does-a-medical-laboratory-technician-do.https://www.ncc.edu/programsandcourses/%20academic_departments/alliedhealthsciences/medicallabtech/jobdescription.shtml.

– Clinical Lab Technician. Accessed on 12-2-2023. https://www.rasmussen.edu/degrees/health-sciences/blog/what-is-clinical-lab-technician/.

– Chemical Lab Technician Responsibilities and Duties. Accessed on 13-2-2023. https://www.greatsampleresume.com/job-responsibilities/chemistry/lab-technician.

– Wastewater Laboratory Technician. Accessed on 13-2-2023. http://joplinmo.org/DocumentCenter/View/10269/Wastewater-Laboratory-Technician.

– Environmental Lab Technician. Accessed on 14-2-2023. https://www.environmentalscience.org/career/lab-technician.

– Environmental Laboratory Technologist. Accessed on 14-2-2023. https://www.nrep.org/blog/environmental-lab-technician-duties.

– Agricultural and Food Science Technician. Accessed on 15-2-2023. https://www.careerexplorer.com/careers/agricultural-and-food-science-technician/.

– Scientific Laboratory Technician. Accessed on 15-2-2023. https://www.prospects.ac.uk/job-profiles/scientific-laboratory-technician.

https://doi.org/10.1515/9783111191492-014

– Laboratory Information Management Systems (LIMS) Information. Accessed on 27-2-2023. https://www.globalspec.com/learnmore/industrial_engineering_software/engineering_scientific_software/laboratory_information_management_systems_lims.

– MSC-LIMS Product Summary. Accessed on 28-2-2023. http://www.msc-lims.com/lims/benefits.html.

– Clinical Technician Job Description. Accessed on 2-3-2023. https://www.betterteam.com/clinical-technician-job-description.

– LABORATORY INTERNAL CHAIN OF CUSTODY. Accessed on 5-3-2023. https://www.wada-ama.org/sites/default/files/resources/files/td2021lcoc_final_eng_0.pdf.

– Nuclear Magnetic Resonance Spectroscopy (NMR Spectroscopy), 2014. Accessed on 7-3-2023. https://itnsnal.net/2014/10/nuclear-magnetic-resonance-spectroscopy/.

– UV-Vis Spectroscopy: Principle, Strengths and Limitations and Applications, 2023. Accessed on 9-3-2023. https://www.technologynetworks.com/analysis/articles/uv-vis-spectroscopy-principle-strengths-and-limitations-and-applications-349865.

– IR Spectroscopy, 2021. Accessed on 10-3-2023. https://byjus.com/chemistry/infrared-spectroscopy/.

– Lab Safety Manual: Working with Hazardous Materials, 2012. Accessed on 13-3-2023. https://www.hampshire.edu/lab-safety-manual-working-hazardous-materials.

– Hazard Communication Program Manual, 2017. Accessed on 15-3-2023. https://ehs.wiscweb.wisc.edu/wp-content/uploads/sites/25/2017/01/HazardCommunicationPolicy.pdf.

– WHO Training Manual, GLP, 2008. Accessed on 18-3-2023. https://proto.ufsc.br/files/2012/03/glp_trainee_green.pdf.

– 8 Good Laboratory Practice Examples, 2023: Accessed on 18-3-2023. https://blog.biobide.com/8-good-laboratory-practice-examples.

– What Is a LIMS? 2019.Accessed on 21-3-2023. https://www.thermofisher.com/blog/connectedlab/what-is-a-lims/.

– Laboratory Information Management Software, 2023. Accessed on 21-3-2023. https://solution4labs.com/en/laboratory-information-management-software-lims-101.

– What Is LIMS & Its Types? – For Medical Labs, 2023. Accessed on 21-3-2023. https://blog.creliohealth.com/what-is-lims/.

– Working with Laboratory Glassware, 2009. Accessed on 28-3-2023. https://www.labmanager.com/lab-health-and-safety/working-with-laboratory-glassware-20359.

– Essentials of Laboratory First Aid, 2014. Accessed on 28-3-2023. https://lab-training.com/essentials-laboratory-first-aid/.

– Validated SaaS LIMS, 2020. Accessed on 31-3-2023. https://www.labware.com/lims/saas?utm_term=saas%20lims&utm_campaign=LW_ENG_NB+Search_Products&utm_source=adwords&utm_medium=ppc&hsa_acc=1902638030&hsa_cam=12087167946&hsa_grp=121695015292&hsa_ad=491896642079&hsa_src=g&hsa_tgt=kwd-307264681090&hsa_kw=saas%2.

– What Are the Different Types of Laboratory Glassware? 2014. Accessed on 1-4-2023. https://www.labmate-online.com/news/laboratory-products/3/breaking-news/what-are-the-different-types-of-laboratory-glassware/32653.

– Chemistry Laboratory Glassware Gallery, 2019. Accessed on 8-4-2023. https://www.thoughtco.com/chemistry-laboratory-glassware-gallery-4054177.

– Laboratory Glassware Disposal Guidelines, 2022. Accessed on 8-4-2023. https://ehs.fiu.edu/_assets/docs/lab-safety/lab-glassware-disposal-guidelines.pdf.

– Laboratory Glassware Disposal Guidelines, 2022. Accessed on 8-4-2023. https://ehs.fiu.edu/_assets/docs/lab-safety/lab-glassware-disposal-guidelines.pdf.

Index

https://doi.org/10.1515/9783111191492-015

www.ingramcontent.com/pod-product-compliance
Lightning Source LLC
Chambersburg PA
CBHW081543220326
41598CB00036B/6538